What Role Can Land-Based, Multi-Domain Anti-Access/Area Denial Forces Play in Deterring or Defeating Aggression?

Timothy M. Bonds, Joel B. Predd, Timothy R. Heath,
Michael S. Chase, Michael Johnson, Michael J. Lostumbo,
James Bonomo, Muharrem Mane, Paul S. Steinberg

Prepared for the United States Army
Approved for public release; distribution unlimited

For more information on this publication, visit www.rand.org/t/RR1820

Library of Congress Cataloging-in-Publication Data is available for this publication.
ISBN: 978-0-8330-9746-0

Published by the RAND Corporation, Santa Monica, Calif.
© Copyright 2017 RAND Corporation
RAND® is a registered trademark.

Limited Print and Electronic Distribution Rights

This document and trademark(s) contained herein are protected by law. This representation of RAND intellectual property is provided for noncommercial use only. Unauthorized posting of this publication online is prohibited. Permission is given to duplicate this document for personal use only, as long as it is unaltered and complete. Permission is required from RAND to reproduce, or reuse in another form, any of its research documents for commercial use. For information on reprint and linking permissions, please visit www.rand.org/pubs/permissions.

The RAND Corporation is a research organization that develops solutions to public policy challenges to help make communities throughout the world safer and more secure, healthier and more prosperous. RAND is nonprofit, nonpartisan, and committed to the public interest.

RAND's publications do not necessarily reflect the opinions of its research clients and sponsors.

Support RAND
Make a tax-deductible charitable contribution at
www.rand.org/giving/contribute

www.rand.org

Preface

This report documents the results of the RAND Arroyo Center project entitled "Strategic and Service Implications of Developing and Deploying Land-Based A2AD Systems by the U.S. Army." The objective of this study was to examine how fielding land-based anti-access/area denial (A2/AD) capabilities would affect regional, political, economic, and military dimensions of relations in key regions, as well as what would need to be done to establish appropriate force structure, doctrine, concept of operations, and other requirements to support counter-A2/AD strategy. This report examines concepts for employing land-based, multi-domain A2/AD forces to deter or defeat aggression in the western Pacific, European littoral areas, and the Persian Gulf.

This research was sponsored by the Headquarters, Department of the Army, Deputy Chief of Staff, G-8, Army Quadrennial Defense Review Office, and conducted within the RAND Arroyo Center's Strategy, Doctrine, and Resources Program. RAND Arroyo Center, part of the RAND Corporation, is a federally funded research and development center sponsored by the United States Army.

The Project Unique Identification Code (PUIC) for the project that produced this document is HQD156917.

Contents

Preface iii
Summary ix
Acknowledgments xxiii

CHAPTER ONE
Introduction 1
Background 1
Study Objective and Motivation 10
Organization of This Document 12

CHAPTER TWO
China in the Western Pacific: Core Interests and Strategic Intentions 15
China's Core Interests 16
China's Strategic Intentions 17
Disputes with the United States 18
Sovereignty Disputes with Japan 21
Sovereignty Disputes with the Philippines 25
Sovereignty Disputes with Taiwan 32
Current Chinese Focus on Resolving Sovereignty Disputes 33

CHAPTER THREE
China-Japan Relationship from Japan's Standpoint 35
Context: U.S. Relationship with Japan 35
Potential China-Japan Conflict Scenarios 36

Addressing the China Threat from Japan's Standpoint..................... 40
What the U.S. Army Can Do to Help the Japanese GSDF Help Itself... 43

CHAPTER FOUR
China-Philippines Relationship from the Philippines' Standpoint.... 45
Context: U.S. Relationship with the Philippines............................ 45
Potential China-Philippines Conflict Scenarios............................. 47
Addressing the China Threat from the Philippines' Standpoint............ 51
What the U.S. Military Could Do to Help the Philippines Help Itself... 56

CHAPTER FIVE
China-Taiwan Relationship from Taiwan's Standpoint.................. 59
Context: U.S. Relationship with Taiwan..................................... 59
Potential China-Taiwan Conflict Scenarios................................. 62
Addressing the China Threat from Taiwan's Standpoint.................... 65
What the U.S. Army Could Do to Help Taiwan Help Itself.............. 68

CHAPTER SIX
The Growing Chinese A2/AD Threat and Blue A2/AD Strategies
  and Operational Concepts to Counter It............................ 71
Expected Growth in China's National Power and Its Impact on Its
  Military Capability 72
The Chinese A2/AD Threat... 73
What Approaches Can the United States and Its Allies and Partners
  Take to Defeat Aggression Shielded by A2/AD Forces?................ 75
Imposing Blue A2/AD Challenges to Deter or Defeat Aggression......... 77
Imposing Blue A2/AD—Illustrative China Cases......................... 82
The Russian A2/AD Threat... 91
Defending the Persian Gulf... 95

CHAPTER SEVEN
Potential Roles for U.S. Land-Based Fires in Joint Missions........... 99
Constraints on U.S. Land-Based Fires 99
Potential Army Roles in Anti-Ship Operations.......................... 106
Potential Army Roles in Long-Range Ground Strike Operations........ 117
Potential U.S. Army Roles in Defending Against Low-Altitude Air
  and Cruise Missile Attacks... 131

Results of Ground-Based A2/AD Mission Assessment...................... 134
Estimates of the Costs to Field Land-Based Missile Systems 136
Near-Term Force Structure Options for Ground-Based
  Multi-Domain Fires 143
Future Force Structure Options for Ground-Based
  Multi-Domain Fires 147

CHAPTER EIGHT
Recommendations, Open Questions, and Next Steps 149
Areas for Further Analysis and Development 151

Figures and Tables.. 153
Abbreviations... 155
References 159

Summary

Context

China is seeking to build a regional and international order that elevates its status and influence, something that provides a primary strategic driver for China's pursuit of control over Taiwan and the East and South China Seas. Given Chinese territorial ambitions and China's aspirations for greater regional influence, the highest potential for crisis and conflict in the western Pacific lies in disputes between China and its neighbors that escalate into armed conflict. Because of U.S. security commitments to its allies in the region, a conflict involving China and an ally would almost unavoidably involve the United States at some level.

Chinese military writing is placing increased emphasis on how to counter U.S. capabilities coming to the aid of the United States' western-Pacific allies. The People's Liberation Army has embarked on an ambitious modernization program that is steadily fielding a formidable anti-access/area denial (A2/AD) force—including A2 capabilities that limit the ability of opposing forces to enter an operational area and AD capabilities that degrade the ability of opposing air and naval forces to operate or maneuver freely.

Russia also poses an A2/AD challenge to the United States and its North Atlantic Treaty Organization (NATO) allies. In 60 hours or less, Russian forces are capable of overrunning forces currently postured in the Baltics.[1] NATO forces racing to reinforce the Baltic states must pass within range of Russian A2/AD forces, most notably those in Kaliningrad. Operating under this A2/AD umbrella, Russian naval and air forces could mount attacks and amphibious operations in the rear areas of Estonia and Latvia, seize Gotland and other strategic islands, and interdict sea traffic from Stockholm to NATO forces in Riga and Tallinn.

To address these challenges, RAND Arroyo Center analysts examined the role that friendly land-based, multi-domain forces can play in deterring or defeating aggression shielded by an A2/AD umbrella. The focus is primarily on China, but the report also examines the use of multi-domain forces to defend against Russia and, briefly, against Iran. The report also seeks to better understand the objectives, interests, goals, and capabilities of China and U.S. allies in the western Pacific to set the strategic context for improving the defenses of U.S. allies.

Key Findings and Conclusions

The United States and its allies can adopt several strategies to counter an adversary's aggression when shielded by A2/AD capabilities. The traditional approach is to establish regional bases from which to operate land, air, and maritime forces projected from the United States. Some of the deployed forces are then assigned to defend these bases against attack by enemy A2/AD systems while simultaneously degrading those same A2/AD capabilities. As an example, in Operations Desert Shield and Desert Storm, the United States and its allies deployed forces to regional bases to reverse Iraq's invasion and occupation of Kuwait. U.S. air, land, and sea forces defended these bases while conducting air and missile attacks against Iraqi air defense and ballistic missile units.[2]

---

[1] David A. Shlapak and Michael Johnson, *Reinforcing Deterrence on NATO's Eastern Flank: Wargaming the Defense of the Baltics*, Santa Monica, Calif.: RAND Corporation, RR-1253-A, 2016.

[2] For a description of U.S. operations to attack Iraqi air defenses and ballistic missiles in Operation Desert Storm—an early campaign to counter A2/AD systems—see James Winnefeld,

The United States took a similar approach in Operation Iraqi Freedom a decade later by establishing and defending bases in Saudi Arabia, Qatar, and Jordan against Iraqi air and ballistic missile attacks while attacking Iraqi air defenses and ballistic missile units.[3]

It would be very difficult to successfully employ a similar strategy against peer or near-peer military forces.[4] For example, if the United States were to assist Japan in defending the Ryukyu Islands from an attack by China, regional airports and seaports would be well within range of Chinese forces estimated to possess hundreds of ballistic missiles, as well as within the range of modern aircraft and cruise missiles. Taiwan faces an even greater challenge in defeating a Chinese invasion. In addition to the air and missile forces that could threaten the Ryukyus, Taiwan is within range of Chinese rocket artillery forces that could number in the hundreds of firing units and employ cluster munitions against aircraft, personnel, and the logistics infrastructure at fixed bases. Even expeditionary bases would be vulnerable once discovered. NATO allies and partners could face an even greater challenge against a Russian attack in the Baltics. In each case, it is not likely practical to defend against cheap and plentiful artillery rockets with expensive defensive missiles that are available in limited numbers.

Preston Niblack, and Dana J. Johnson, *A League of Airmen: U.S. Air Power in the Gulf War*, Santa Monica, Calif.: RAND Corporation, MR-343-AF, 1994.

[3] Several reports describe operations against Iraqi air defenses and ballistic missiles in Operation Iraq Freedom and an account of U.S. air and missile defense operations. See 32nd Army Air and Missile Defense Command, *Operation Iraqi Freedom Theater Air and Missile Defense History*, Fort Bliss, Tex., September 2003; T. Michael Mosely, *Operation Iraqi Freedom—By the Numbers: Assessment and Analysis Division*, Shaw Air Force Base, S.C.: U.S. Central Command Air Forces, April 30, 2003; and Gregory Fontenot, E. J. Degen, and Dave Tohn, *On Point—The United States Army in Operation Iraqi Freedom*, Fort Leavenworth, Kan.: Combat Studies Institute Press, 2004.

[4] For an analysis of the damage that cruise and ballistic missiles might inflict on air bases, see John Stillion and David T Orletsky, *Airbase Vulnerability to Conventional Cruise-Missile and Ballistic-Missile Attack*, Santa Monica, Calif.: RAND Corporation, MR-1028-AF, 1999; and David A. Shlapak, David T. Orletsky, Toy I. Reid, Murray Scot Tanner, and Barry Wilson, *A Question of Balance Political Context and Military Aspects of the China-Taiwan Dispute*, Santa Monica, Calif.: RAND Corporation, MG-888-SRF, 2009.

In these cases, U.S. allies, with support from the United States, should consider ways to *impose A2/AD challenges on enemies*.[5] This would allow allied forces to contest maritime areas without exposing U.S. forces to easy attack. Such Blue A2/AD capabilities might be a particularly effective way to raise the costs for aggression. **In particular, U.S. allies and partners could field a mix of anti-ship, anti-aircraft, and surface-to-surface missiles to impose the same problems on adversaries threatening them with attack over water. The U.S. joint force could provide support—and potentially reinforcements—to its allies.**

A combination of A2/AD capabilities in the hands of U.S. allies and partners and the U.S. joint forces might be especially useful under the following conditions:

- The magnitude of a potential adversary's A2/AD forces renders theater bases and forces vulnerable to preemptive strikes, threatening the survivability of allied air and sea forces and compelling U.S. forces to begin operations outside theater.
- Warning time is short, which means that an act of aggression might begin before U.S. air and sea forces can position themselves inside enemy A2/AD defenses, requiring forces to break through to deploy to theater.
- U.S. air and sea forces are already heavily committed in other operations—for example, to deterring or defeating Russian aggression in the Baltics when a conflict erupts elsewhere in the world.
- The United States would prefer not to project power into a region in the midst of a crisis, especially if doing so might escalate a conflict when tensions are already very high.

---

[5] Previous RAND work has described the geography of Blue A2/AD in the western Pacific, as well as the strategic rationale for building partner capacity. See Terrence K. Kelly, Anthony Atler, Todd Nichols, and Lloyd Thrall, *Employing Land-Based Anti-Ship Missiles in the Western Pacific*, Santa Monica, Calif.: RAND Corporation, TR-1321-A, 2013; and Terrence K. Kelly, David C. Gompert, and Duncan Long, *Smarter Power, Stronger Partners: Exploiting U.S. Advantages to Prevent Aggression*, Santa Monica, Calif.: RAND Corporation, RR-1359-A, 2016.

- U.S. allies and partners could deploy the majority of these forces in their own defense.

Such defensive concepts might give the U.S. military graduated options to respond to threats. First and foremost, Blue A2/AD concepts shift the primary responsibility for defense to U.S. allies and partners. The systems involved are affordable by U.S. allies and can be dispersed, hidden, and moved to better survive attack than can fixed facilities. Rather than requiring the U.S. military to build and employ these systems for its allies, the United States could instead field capabilities with or in support of its allies. In peacetime, the United States might begin by working with key allies and partners to develop new operational concepts, modernize equipment, and participate in combined training and exercise events, such as Pacific Pathways. In times of crisis, the United States could increase its support to allies by providing key enablers, such as robust communications; intelligence, surveillance, and reconnaissance (ISR); and targeting. If a conflict begins or appears imminent, the United States could help allies defend key facilities, bases, or strategic locations. Finally, the United States could reinforce allies to defeat aggression.

Anti-Ship Missions

We examined three different types of anti-ship missions: (1) tactical coastal defenses to defeat an amphibious landing, (2) maritime interdiction to deny the operations of enemy naval forces in a broad area, and (3) enforcing blockades of specific ports.

The Pacific theater clearly demands that the United States and its allies and partners possess anti-ship capabilities. Existing U.S. naval and air forces bear the brunt of the responsibility and compose the existing and planned capacity and capability to conduct anti-ship missions. Thus, the need for ground-based U.S. anti-ship forces is likely limited in the Pacific. However, Taiwan, the Philippines, and perhaps Japan face challenges in defending their territories against potential threats from China and, in the case of Japan, Russia.

The A2/AD threat in the Baltics, combined with Russia's own air and naval capabilities, should make anti-ship strike operations a major mission for NATO nations and partners, such as Sweden and Finland. Deploying anti-ship capabilities faster or sooner may be important in some cases, such as if hostile A2/AD or short warning could limit the ability to deploy allied or U.S. air and naval forces.

This implies two potential roles for the U.S. Army: (1) help allies and partners build coastal defense capabilities and concepts and (2) develop a small and specialized Army unit with longer-range anti-ship capabilities for precrisis deployment or rapid deployment once a crisis or conflict has begun. This capability might be most advantageous in the Baltics, but would have to compete for funding with many other priorities in an already severely stretched Army budget.

Surface-to-Surface Strike

Surface-to-surface strike is a critical capability to conduct counterbattery fires against enemy anti-ship missile (ASM) batteries, long-range anti-aircraft missile systems, and adversary air and sea bases. Current U.S. Army strike systems—for example, the Army Tactical Missile System (ATACMS) with its 300-kilometer range—have value today if they are based on allied soil within range of their targets. Other U.S. ground-based strike systems would be even more valuable if their ranges were increased to the 499-kilometer limit imposed by the Intermediate-Range Nuclear Forces (INF) Treaty.

There might be value in building a very long-range ballistic missile force able to hold adversary air bases, command and control nodes, and other sites at risk of a rapid attack. Furthermore, there is a joint gap in the capability to strike targets in less than ten minutes from standoff ranges that exceed 500 kilometers, and there is a plausible, but untested, hypothesis that a capability to strike targets from 500 to 1,500 kilometers with such responsiveness could have disproportionate effects at the campaign level. However, existing analyses have not validated a joint demand for long-range surface-to-surface missiles, and the INF Treaty precludes their development beyond 500 kilometers.

Short-Range Air and Cruise Missile Defense

The demands for air and cruise missile defense are great in the western Pacific, owing both to the threat posed by the large Chinese investment in cruise missiles and to emerging joint operational concepts that geographically disperse air bases. The same is likely true to an even greater extent in Eastern Europe in potential operations against Russia. Air defenses play a key role in countering low-altitude aircraft and cruise missiles today, but their capability and capacity will become increasingly strained as the threat grows.

Specifically, there is a gap in the ability of current and planned joint forces to provide high volumes of fire to counter aircraft and cruise missiles at short ranges. Moreover, in wartime, other missions will compete for Air Force (e.g., fighter aircraft) and Navy assets (e.g., Aegis destroyers), thus limiting those services' availability for low-altitude air and cruise missile defense. Systems that can provide high volumes of fire and tactical agility to avoid destruction and to accommodate responsive deployments would be very valuable.

Fortunately, the Army has systems in development that are programmed to appear in 2018—notably, the Indirect Fire Protection Capability–Increment 2 (IFPC-2)—that could squarely address this gap with lighter-weight, more quickly and easily transported launchers and deep magazines potentially enabled by relatively cheap, off-the-shelf AIM-9 interceptors. The IFPC-2 also offers a land-based alternative for allies and partners to defend themselves and enables a more *offensive* approach of thinking about air and cruise missile defense by providing a way to shape enemy weapon-target pairing choices.

However, the Army plans for most IFPC-2s to go to the reserve, not active forces, which may limit the availability of these forces for routine deployments and early entry in warfighting operations. Also, unless the AIM-9s needed for these systems are provided from existing joint stocks, current Army budgets would severely limit the numbers of these munitions that could be purchased. If these systems could provide a valuable option for U.S. strategy in supporting the defense of U.S. allies, then U.S. Department of Defense budgets should provide the funds necessary to purchase the needed systems and munitions and fund their operations.

Table S.1 summarizes our findings and conclusions about ground-based A2/AD missions, including *anti-ship missions* (tactical coastal defense, theater maritime interdiction, and lethal blockade

Table S.1
Ground-Based A2/AD Missions—Findings and Conclusions

Mission	Potential Adversary: China Allies: Japan and the Philippines	Potential Adversary: China Partner: Taiwan	Potential Adversary: Russia Ally: NATO	Potential Adversary: Iran Partners: Persian Gulf Nations
Anti-Ship Operations				
Tactical coastal defense	• Demand high • Potential U.S. policy constraints • Natural mission for Japan and the Philippines	• Demand very high • Policy prohibits peacetime U.S. presence • Natural mission for Taiwan	• Demand very high • Red A2/AD limits NATO sea and air forces • Natural NATO and partner mission	• Demand moderate • U.S. joint anti-ship capability high • Mission for Persian Gulf partners
Theater maritime interdiction	• Demand high • Joint anti-ship capability high • Natural mission for Japan and the Philippines • United States could assist and reinforce	• Demand very high • Joint anti-ship capability high • Natural mission for Taiwan • United States might assist • ASMs could arrive faster	• Demand very high • Red A2/AD limits NATO sea and air forces • Natural NATO and partner mission • United States could assist and reinforce	• Demand moderate • U.S. joint anti-ship capability high • Natural mission for Persian Gulf partners • United States could assist and reinforce
Lethal blockade enforcement	• Demand narrow and rare; blockades are enforced by naval forces in most scenarios. Lethal enforcement by land-based ASMs requires multiple unusual assumptions about the conflict and the availability of naval forces.			
Surface-to-Surface Long-Range Strike Operations				
<500 km strike	• Existing and improved systems useful for suppressing or destroying enemy air defenses, conducting counterbattery, and striking airfields and other high-value targets.			
>500 km strike	• Very long-range sniper concept could offer responsiveness and ability to penetrate enemy defenses, but analysis is needed to examine hypothesized effects • Demand plausible but not established • Operational benefit would be required to justify withdrawal from INF Treaty			
Strike in denied environments	• Demand plausible but not established • A2 environment would inherently limit employment			

Table S.1—Continued

Mission	Potential Adversary: China Allies: Japan and the Philippines	Partner: Taiwan	Potential Adversary: Russia Ally: NATO	Potential Adversary: Iran Partners: Persian Gulf Nations
	Short-Range Air and Cruise Missile Defense			
Point defenses	• Demand very high because of emerging threats and distributed basing • Joint capability and capacity gaps • Army's emerging IFPC-2 could address gap if force designs and force structure accommodate demands for responsiveness and survivability • Munition supply represents significant expense			

enforcement), *surface-to-surface missions* (long-range strike and fires in denied environments), and *short-range air and cruise missile defense missions*. Each of these missions is considered in defense of U.S. allies and friends: Japan, the Philippines, and Taiwan against potential Chinese aggression; NATO Baltic and Black Sea allies against potential Russian aggression; and Persian Gulf allies against potential Iranian aggression. The first two anti-ship missions are broken out by adversary and ally, while the third anti-ship mission and the other two missions apply across allies and adversaries.

Recommendations

A fundamental choice for the United States is whether to focus on building and supporting allied defensive concepts or on employing primarily U.S.-only concepts on behalf of allies. That is, should the United States act with and in support of its allies, or should it act for them? Ground-based concepts offer an affordable way for allies and partners to take the primary responsibility for their own defense, and they offer an attractive way for the United States to build capabilities to operate in support of allies.

Based on the findings and conclusions above, the Army should organize and field a prototype multi-domain fires battalion to develop, test, and exercise joint and combined defensive concepts discussed in the report. Table S.2 provides a set of recommendations.

To start as quickly and inexpensively as possible, long-range ISR and targeting capabilities should be provided by existing U.S. Navy, U.S. Air Force, and allied systems.

For the anti-ship role, an initial capability could be established by building a combined battalion incorporating existing ASM batteries operated by selected allies, such as Poland and Japan. Additional allies could seek to join as they develop the requisite capabilities. If current development programs succeed in building versions of the ATACMS or the Guided Multiple Launch Rocket System (GMLRS) with a terminal guidance package for anti-ship operations, the multi-domain battalion should incorporate a U.S. Army anti-ship battery capable of operating them. This is a preferable, longer-term option that provides more flexibility for employing both ground-attack and anti-ship versions of the ATACMS from the same launchers. Using systems already in the U.S. inventory would also simplify logistics and support operations.

Table S.2
Recommendations for Prototype Multi-Domain Fires Battalion for Joint Concept Development and Experimentation

Multi-Domain Capability	Recommendations for the Army
Long-range ISR and targeting	• Rely on U.S. Navy, U.S. Air Force, and allied forces and systems
Surface-to-surface capabilities	• Assign the High-Mobility Artillery Rocket System (HIMARS) battalion equipped with ATACMS and the Guided Multiple Launch Rocket System (GMLRS)

Anti-ship capabilities	• Assign ASM battery
	  – Begin with allied batteries in Combined U.S.-Allied multi-domain battalions
	  – Develop anti-ship version of ATACMS for U.S. HIMARS battery

Air and cruise missile defense capabilities	• Develop deployable IFPC-2 minimum engagement packages (MEPs) • Assign battery of three IFPC-2 MEPs into multi-domain battalion

An existing High-Mobility Artillery Rocket System (HIMARS) battery should be assigned to provide surface-to-surface fires. These fires could contribute to the suppression of enemy air defenses missions, disrupt operations at enemy air bases, and attack amphibious forces that have landed on friendly territory. This battery could be equipped with both ATACMS and a shorter-range GMLRS. Improved long-range missiles, such as the 499-kilometer Long-Range Precision Fires system, could extend the range of this battery when fielded. If an anti-ship version of the ATACMS or GMLRS is developed, then the anti-ship and surface-to-surface batteries could be interchangeable and assigned targets more flexibly—combining the firepower that could be delivered against either target set as operational priorities dictate.

Finally, a short-range air and cruise missile defense battery could be assigned from the forces being formed to operate IFPC-2. To provide a capability that can operate in small numbers, the U.S. Army should develop and deploy minimum engagement packages (MEPs) for exercises and demonstrations with allies and partners. Three platoon-sized MEPs could be assigned as the air and cruise missile defense battery.

If the prototype battalion proves useful, additional multi-domain fires battalions could also be formed. These battalions could represent new force structure or assignments of existing HIMARS/Multiple Launch Rocket System (MLRS) batteries and IFPC-2 platoons. In addition, future battalions could be equipped with Long-Range ASMs or longer-ranged surface-to-surface weapons.

After an initial set of joint operating concepts has been developed, the Army should work with key allies and partners to build combined concepts and tactics, techniques, and procedures. Allies and partners could include Japan, the Philippines, and Taiwan in the Asia-Pacific region; NATO nations (Poland, for example, already fields the Norway Naval Strike Missile in coastal defense squadrons); Sweden; Finland; and Gulf Cooperation Council members in the Persian Gulf. The United States could plan and host joint and combined exercise events to refine these concepts and provide continuing training for allies and partners.

Areas for Further Analysis and Development

This analysis leaves open several questions that warrant further analysis and development.

Force Size and Mix

Our recommendations for a prototype multi-domain fires battalion are designed to enable joint concept development and experimentation. Consequently, the mix of surface-to-surface, anti-ship, and air and cruise missile defense capabilities are balanced to enable experimentation rather than meet particular operational objectives. In practice, the size and mix of such capabilities will vary based on the objectives and missions in a particular scenario. A deeper examination of the size and mix of the multi-domain fires battalion is warranted. In addition, the large "advise and assist" role of the Cross-Domain Fires Battalion needs to be considered, because that will require more officers, senior enlisted, and communications capability than normal HIMARS and MLRS units.

Mission Scope

This work focused primarily on surface-to-surface, anti-ship, and surface-to-air capabilities without examining potential missions for a multi-domain task force in the space and cyber domains or multi-domain battle roles that employ maneuver, force protection, aviation, or sustainment functions. Additionally, other multi-domain concepts of operation are worthy of consideration, such as potentially establishing land-based defense of carrier battle groups and defending land bases from asymmetric mortar or special operations forces attacks, among others. An examination of the broader mission scope of a multi-domain battle task force is an important area for future work.

Joint Wargaming

We analyze the potential demands associated with allied and U.S. ground forces organized and equipped with multi-domain fires. One way to test the value proposition of such a unit is through a series of formally adjudicated wargames, where active Blue and Red teams are afforded the choice of employing or countering the proposed multi-domain fires battalion. Such wargames would be a useful analytic component to the proposed joint concept development and experimentation and enable further refinement of these ideas.[6]

Fire Support Coordination

The Army does have cross-domain coordination experience in air and missile defense, but the coordination of fires between the Joint Force Land Component Commander and the Joint Force Maritime Component Commander is new. This will require new doctrine and new coordination measures with the maritime components.

---

[6] For an example of a wargame, built on detailed quantitative and qualitative analyses, see David A. Shlapak and Michael Johnson, *Reinforcing Deterrence on NATO's Eastern Flank: Wargaming the Defense of the Baltics*, Santa Monica, Calif.: RAND Corporation, RR-1253-A, 2016.

Acknowledgments

The authors would like to thank MG Steven Grove and Tim Muchmore (G-8/QDRO), who sponsored this work. In addition, we would like to thank RAND colleagues Michael Hansen, Sally Sleeper, Mike Decker, and David Luckey for their excellent reviews, comments, and other assistance. Finally, we would like to thank Rebecca Fowler for her assistance in producing this document.

CHAPTER ONE

Introduction

Background

The Importance of the Global Commons to the United States

The United States has strongly upheld what has come to be called the *global commons*—the freedom of navigation and the right to transit or use international waterways and the airspace above them—a phrase that has shown up in key policy documents since the early 2000s. For example, the 2004 National Military Strategy stated: "The Department [of Defense] must work to secure strategic access to key regions, lines of communication and the 'global commons' of international waters, airspace, space and cyberspace."[1]

The 2008 National Defense Strategy expanded on this idea, describing the motivation and means for maintaining access to these global commons for all:

> For more than sixty years, the United States has secured the global commons for the benefit of all. Global prosperity is contingent on the free flow of ideas, goods, and services. . . . The development and proliferation of anti-access technologies and tactics threatens to undermine this. . . .
>
> The United States requires freedom of action in the global commons and strategic access to important regions of the world to

---

[1] Joint Chiefs of Staff, *The National Military Strategy of the United States of America: A Strategy for Today; A Vision for Tomorrow*, Washington, D.C.: Office of the Chairman of the Joint Chiefs of Staff, 2004.

meet our national security needs. The well-being of the global economy is contingent on ready access to energy resources. . . . The United States will continue to foster access to and flow of energy resources vital to the world economy.

We will continue . . . developing a more relevant and flexible forward network of capabilities and arrangements with allies and partners to ensure strategic access.[2]

Securing the Global Commons from Threats Posed by Anti-Access/Area Denial

But U.S. access to these global commons might be under threat, as indicated in the 2010 National Security Strategy (NSS), which states:

We must also safeguard the sea, air, and space domains from those who would deny access or use them for hostile purposes. This includes keeping strategic straits and vital sea lanes open, improving the early detection of emerging maritime threats, denying adversaries hostile use of the air domain, and ensuring the responsible use of space.[3]

Who might threaten the peaceful access to international air and maritime areas by the United States or its allies and partners? For what purposes? And what can and should the U.S. military be prepared to do in response?

In the 2012 Defense Strategic Guidance, the Secretary of Defense, Leon Panetta, named China and Iran as potential threats and applied the term *anti-access and area denial* (A2/AD) to an operating concept to control air and maritime areas. He also named the Joint Operational Access Concept (JOAC) as a U.S. Department of Defense (DoD) response:

States such as China and Iran will continue to pursue asymmetric means to counter our power projection capabilities. . . . Accord-

---

[2] U.S. Department of Defense, *National Defense Strategy*, Washington, D.C., June 2008.

[3] Barack Obama, *National Security Strategy*, Washington, D.C.: White House, May 2010.

ingly, the U.S. military will invest as required to ensure its ability to operate effectively in anti-access and area denial (A2/AD) environments. This will include implementing the Joint Operational Access Concept.[4]

The 2014 Quadrennial Defense Review gave one potential rationale for China's interest in keeping out U.S. power projection forces:

> As nations in the region continue to develop their military and security capabilities, there is greater risk that tensions over long-standing sovereignty disputes or claims to natural resources will spur disruptive competition or erupt into conflict, reversing the trends of rising regional peace, stability, and prosperity. In particular, the rapid pace and comprehensive scope of China's military modernization continues, combined with a relative lack of transparency and openness from China's leaders regarding both military capabilities and intentions.[5]

This was followed the next year by these words in the 2015 NSS:

> [T]ensions in the East and South China Seas are reminders of the risks of escalation.
>
> . . . [W]e will manage competition from a position of strength while insisting that China uphold international rules and norms on issues ranging from maritime security to trade and human rights. We will closely monitor China's military modernization and expanding presence in Asia, while seeking ways to reduce the risk of misunderstanding or miscalculation.[6]

And these comments from the NSS were echoed in the 2015 National Military Strategy:

---

[4] U.S. Department of Defense, *Sustaining U.S. Global Leadership; Priorities for 21st Century Defense*, Washington, D.C., January 2012b, pp. 4–5.

[5] U.S. Department of Defense, *Quadrennial Defense Review 2014*, Washington, D.C., 2014b.

[6] Barack Obama, *National Security Strategy*, Washington, D.C.: White House, February 2015.

China's actions are adding tension to the Asia-Pacific region. For example, its claims to nearly the entire South China Sea are inconsistent with international law. The international community continues to call on China to settle such issues cooperatively and without coercion. China has responded with aggressive land reclamation efforts that will allow it to position military forces astride vital international sea lanes.[7]

Concerns about aggression and A2/AD extend beyond the Asia-Pacific region. Recent Russian aggression in Ukraine has raised the concern that U.S. allies and friends, including North Atlantic Treaty Organization (NATO) Baltic Sea allies, might also be threatened. If Russia were to attack the Baltic states, it would most likely use ground forces to quickly overrun its borders with Estonia, Latvia, and Lithuania before NATO reinforcements could arrive. Russia has built forces in Kaliningrad to deny access to NATO reinforcements coming by sea or air.

President Barack Obama addressed the Russian threat to the Baltics in his speech in Tallinn, Estonia, in September 2014:

[W]e have made historic progress toward the vision we share—a Europe that is whole and free and at peace.

And yet, as we gather here today, we know that this vision is threatened by Russia's aggression against Ukraine. It is a brazen assault on the territorial integrity of Ukraine—a sovereign and independent European nation. . . .

We have a solemn duty to each other. Article 5 [of the North Atlantic Treaty] is crystal clear: An attack on one is an attack on all. So if, in such a moment, you ever ask again, "who will come to help," you'll know the answer—the NATO Alliance, including the Armed Forces of the United States of America, "right here, [at] present, now!" We'll be here for Estonia. We will be here for

---

[7] Joint Chiefs of Staff, *The National Military Strategy of the United States of America 2015: The United States Military's Contribution to National Security*, Washington, D.C., June 2015.

Latvia. We will be here for Lithuania. You lost your independence once before. With NATO, you will never lose it again.[8]

The 2015 NSS followed up by committing to increased responsiveness and presence to reassure U.S. allies:

> Russia's aggression in Ukraine makes clear that European security and the international rules and norms against territorial aggression cannot be taken for granted. In response, we have led an international effort to support the Ukrainian people as they choose their own future and develop their democracy and economy. We are reassuring our allies by backing our security commitments and increasing responsiveness through training and exercises, as well as a dynamic presence in Central and Eastern Europe to deter further Russian aggression.[9]

The Deputy Secretary General of NATO, Alexander Vershbow, specifically pointed to Russian forces in Kaliningrad as posing a particularly powerful A2/AD threat:

> [F]rom the point of view of the policy-making community in Brussels we're very concerned about the Russian military build-up. The increasing concentration of forces in Kaliningrad and the Black Sea and now in the Eastern Mediterranean does indeed pose some additional challenges that our planners are going to have to take seriously into account as we consider how to live up to the pledge that we have made to defend any Ally against any threat. A lot of experts talk about the so-called anti-access area denial capability that Russia is developing and this indeed will have to be a key factor as we decide what is necessary both to

---

[8] White House, "Remarks by President Obama to the People of Estonia, Nordea Concert Hall, Tallinn, Estonia," Washington, D.C.: Office of the Press Secretary, September 3, 2014a.

[9] Obama, 2015.

defend every Ally and to deter Russia from even thinking about aggressive actions against NATO.[10]

The Deputy Secretary General continued:

> [F]or NATO Allies, one of the most pressing issues coming from this modernisation programme is Russia's A2/AD posture. Anti-Access and Area Denial is the ability, through military capabilities—primarily anti-aircraft and anti-ship missile systems—to prevent Allied forces from moving freely within international waters and airspace to reinforce our own territory. With [Russia's] military build-up on the Crimean peninsula and in the Black Sea—an area of particular worry for Romania— and in Kaliningrad and everywhere from the Barents Sea to the Mediterranean, this is now of serious concern for the Alliance. It is a challenge we must meet in the coming years.[11]

How Can Such A2/AD Threats Be Met?

To address A2/AD threats, DoD has developed the JOAC. The JOAC described the Air-Sea Battle (ASB) concept as a potential solution:

> Recognizing that anti access/area-denial capabilities present a growing challenge to how joint forces operate, the Secretary of Defense directed the Department of the Navy and the Department of the Air Force to develop the Air-Sea Battle Concept. . . . It focuses on ensuring that joint forces will possess the ability to project force as required to preserve and defend U.S. interests well into the future.[12]

---

[10] North Atlantic Treaty Organization, "Joint Press Point—with NATO Deputy Secretary General Ambassador Alexander Vershbow and Italian Undersecretary of Defence, Gioacchino Alfano," transcript, last updated October 20, 2015.

[11] North Atlantic Treaty Organization, "'NATO Post-Warsaw: Strengthening Security in a Tough Neighbourhood'—Speech by NATO Deputy Secretary General Ambassador Alexander Vershbow at the Annual Meeting of Romanian Ambassadors in Bucharest," transcript, August 29, 2016.

[12] U.S. Department of Defense, *Joint Operational Access Concept (JOAC)*, Version 1.0, Washington, D.C., January 17, 2012a, p. 4.

The ASB Office described the problem that A2/AD poses as the following:

> [A]dversary capabilities to deny access and areas to US forces are becoming increasingly advanced and adaptive. These A2/AD capabilities challenge US freedom of action by causing US forces to operate with higher levels of risk and at greater distance from areas of interest. US forces must maintain freedom of action by shaping the A2/AD environment to enable concurrent or follow-on operations.
>
> First, the adversary will initiate military activities with little or no indications or warning. . . . Capabilities such as ballistic and cruise missiles will be used with little warning.
>
> Second . . . forward friendly forces will be in the A2/AD environment at the commencement of hostilities. . . . [S]teady state posture and capabilities of forces must be able to provide an immediate and effective response to adversary A2/AD attacks.
>
> Third, adversaries will attack US and allied territory supporting operations against adversary forces. . . . [D]enying access to US forces requires attacks on bases from which US and its allies are operating, including those on allied or partner territory.[13]

To solve these problems, ASB describes the necessity of attacking all elements of an adversary's A2/AD "kill chain" as soon as a conflict begins, through the entire depth of the battlespace, and across a range of threats:

> Attack-in-depth methodology is based on adversary effects chains, or an adversary's process of finding, fixing, tracking, targeting, engaging, and assessing an attack on US forces . . . with the objec-

---

[13] Air-Sea Battle Office, *AIR-SEA BATTLE: Service Collaboration to Address Anti-Access & Area Denial Challenges*, Washington, D.C.: U.S. Department of Defense, May 2013, p. 3.

tive of disrupting, destroying, or defeating an adversary's A2/AD capabilities.[14]

Due to the nature of A2/AD threats and potentially short indications and warning timelines posed by adversaries, joint forces must be capable of effective offensive operations as soon as conflict begins. . . . The ability to attack and defend through the entire depth of the desired battlespace . . . is critical.[15]

The ASB Concept specifically addresses a range of threats, such as ballistic and cruise missiles, sophisticated integrated air defense systems, [and] anti-ship capabilities from high-tech missiles.[16]

This approach includes striking the A2/AD systems deep, even if located in an opponent's home territory. This is likely to be an extremely difficult challenge, given the long ranges of modern anti-ship missiles (ASMs) and surface-to-air missiles (SAMs); the vast geographic extent that includes cities, forests, mountains, and other hiding sites of Russia, China, Iran, and other potential threat nations; and the great difficulties that U.S. forces have had locating mobile missiles and related systems in Iraq. Even if effective, such an offensive campaign could absorb a significant portion of the U.S. air and sea effort, thus leaving less capacity to deal with counterattacks by adversaries. Worse, attacking targets in China, Russia, or Iran would likely lead to a significant escalation of any conflict, which, in turn, might constrain the President's willingness to employ forces.

A2/AD Works in Both Directions

Given the A2/AD threat to U.S. air and naval forces deploying to defeat an aggressor and the difficulties in directly attacking an adversary's A2/AD systems, the United States has been seeking alternatives. As pointed out in other RAND Arroyo Center work, A2/AD works in both directions. That is, rather than attack Chinese A2/AD systems—

---

[14] Air-Sea Battle Office, 2013, p. 4.
[15] Air-Sea Battle Office, 2013, p. 7.
[16] Air-Sea Battle Office, 2013, pp. 8–9.

including those in China—as the means of introducing air and naval forces that can then interdict Chinese naval forces, it may be simpler and safer to employ ground-based systems to interdict those same naval forces.[17]

This idea has caught on with senior political and military leaders. Representative Randy Forbes, then chairman of the House Seapower Subcommittee, stated:

> The Army's existing expertise in missile defense, rocket and missile systems for offensive precision fires, and partner-capacity building make [the Army] a natural choice to spearhead . . . an integrated web of A2/AD systems among allies [that] would greatly increase the costs of any hostile action in the region.[18]

More recently, the House Armed Services Committee markup of the 2017 National Defense Authorization Act allocated $5 million for the Army to evaluate ground-based ASMs.[19]

In testimony to Congress, Admiral Harry B. Harris, Commander, U.S. Pacific Command (PACOM), commented:

> I believe China seeks hegemony in East Asia. Simple as that. . . . China is clearly militarizing the South China Sea. . . . China's surface-to-air missiles on Woody Island, the new radars on the Cuarteron Reef, the 10,000-foot runway on Subi Reef over here and on Fiery Cross Reef and other places, these are actions that are changing in my opinion the operational landscape in the South China Sea.

---

[17] Terrence K. Kelly, Anthony Atler, Todd Nichols, and Lloyd Thrall, *Employing Land-Based Anti-Ship Missiles in the Western Pacific*, Santa Monica, Calif.: RAND Corporation, TR-1321-A, 2013; Terrence K. Kelly, David C. Gompert, and Duncan Long, *Smarter Power, Stronger Partners*, Vol. 1: *Exploiting U.S. Advantages to Prevent Aggression*, Santa Monica, Calif.: RAND Corporation, RR-1359-A, 2016.

[18] Sydney J. Freedberg, Jr., "Army Should Build Ship-Killer Missiles: Rep. Randy Forbes," *Breaking Defense*, October 12, 2014.

[19] Jane Edwards, "Reports: HASC's 2017 NDAA Calls for Army to Replace Patriot Radar, Assess Land-Based Anti-Ship Missiles," *ExecutiveGov*, April 27, 2016.

We should continue to exercise our rights on the high seas and in the airspace above it. . . . And we should encourage our friends, partners and allies to do the same.[20]

Separately, in comments at the Land Power in the Pacific (LANPAC) symposium, Harris stated:

> Countries like China, Iran, and Russia are challenging our ability to project power ashore, from the sea, through ever-more sophisticated anti-ship missiles. More and more, adversary rocket forces are projecting power over the water in order to protect their control on land. They are also developing land attack missiles and the precision targeting systems that can threaten our facilities ashore. We need systems that enable the Army to project power over water, from shore. Fort Kamehameha hasn't moved an inch since it was built . . . but what we need today is a "Fort HIMARS"—a highly mobile, networked, lethal weapons system with long reach—and if we get this right, the Army will kill the archer instead of dealing with all of its arrows. I believe that the Army should look at ways to use the Paladin and HIMARS [High-Mobility Artillery Rocket System] systems to keep at risk the enemy's Navy[,] . . . not only the enemy's land, which we already do and do well.[21]

Study Objective and Motivation

To address these concerns and concepts, we conducted a series of analyses that examine the role that land-based, multi-domain A2/AD forces can play in deterring or defeating aggression. To begin, to have a complete view of how an adversary might use A2/AD to deny U.S. access to portions of the global commons or use such capabilities to shield

---

[20] Bill Gerts, "Pacific Command: China Seeking to Control South China Sea: Harris Calls for Buildup of U.S. Missiles, Weapons in Region," *Washington Free Beacon*, February 24, 2016.

[21] Harry B. Harris, Jr., "Role of Land Forces in Ensuring Access to Shared Domains," speech presented at the Institute of Land Warfare LANPAC Symposium, Waikiki, May 25, 2016.

attacking forces from a U.S. response, it is important to understand the objectives and intent of potential adversaries. In 2013, RAND Arroyo Center research outlined the effects that could be achieved in the western Pacific with land-based AsMs but caveated that research with the observation that the strategic considerations—e.g., regional political, economic, and security issues—were not considered.[22] Thus, the first motivation for this analysis was to determine how these strategic considerations—the objectives and intent of potential adversaries—will influence decisions and actions in the PACOM, U.S. European Command (EUCOM), and U.S. Central Command areas of responsibility (AORs), as well as the doctrine, organization, training, materiel, leadership, personnel, and facilities (DOTMLPF) implications of fielding AsMs for the Army.

When it comes to potential U.S. adversaries, the two most important are Russia and the People's Republic of China (hereafter, China); therefore, they are the main focus of our analysis, although we make some additional points in the report reflecting Iran and the Persian Gulf. In terms of the interest and objectives of Russia, the answer may be fairly straightforward. Russia views the presence of NATO forces or "infrastructure" on its border as a security threat. Therefore, and especially in the aftermath of Russia's invasion of Crimea and Ukraine, the NATO Baltic nations appear to be at greater risk of attack. This risk might be small, but if Russia ever did choose to take military action, it would need to prevent the Baltics from receiving immediate military assistance from the rest of NATO. Thus, to shield its forces from NATO counterattack, Russia would presumably use the extensive A2/AD capabilities it has established in the Kaliningrad enclave.[23]

The interests and objectives of China are a bit more nuanced and, hence, require a deeper discussion; we also do this for the interests and objectives of its neighbors, including Japan, the Republic of the Philippines (hereafter, the Philippines), and Taiwan.

---

[22] Kelly, Atler, et al., 2013.

[23] See David A. Shlapak and Michael Johnson, *Reinforcing Deterrence on NATO's Eastern Flank: Wargaming the Defense of the Baltics*, Santa Monica, Calif.: RAND Corporation, RR-1253-A, 2016.

After providing an explanation of the objectives of potential adversaries, we focus on how the United States and its allies can address these potential adversaries and on the role the U.S. Army can play in doing so. We describe concepts for the United States and its allies and partners to employ A2/AD defenses—what we refer to as *Blue A2/AD*[24]—to deter or defeat aggression through or over the maritime domain and explore potential Army defensive roles,[25] including providing training and support to allies and coalition partners and providing reinforcements to augment their defenses.

Organization of This Document

In Chapter Two, we present our assessment of the goals and capabilities of China in the western Pacific. We begin by considering China's strategic intentions based on its stated core interests. This includes China's perspectives on disputes with the United States, Japan, the Philippines, and Taiwan. We then present our assessment of how China is currently dealing with those disputes. In Chapter Three, we describe the China-Japan relationship from the standpoint of the Japanese. We include an assessment of potential conflict scenarios, the Japanese security posture, and how the U.S. military can help Japan to help itself. We perform a similar assessment for the Philippines in Chapter Four and Taiwan in Chapter Five.

In Chapter Six, we examine trends in China's national power and evaluate China's growing A2/AD capabilities. We describe concepts for the United States and its allies and partners to employ A2/AD defenses to deter or defeat Chinese aggression through or over the maritime domain. We then make a few related observations for NATO defending against Russia in the Baltics and for the United States and

---

[24] Kelly, Gompert, and Long, 2016.

[25] The *maritime domain*, as defined in Joint Publication 3-32 (*Command and Control for Joint Maritime Operations*, Washington, D.C.: Joint Chiefs of Staff, August 7, 2013), comprises "[t]he oceans, seas, bays, estuaries, islands, coastal areas, and the airspace above these, including the littorals."

its Persian Gulf partners defending against potential Iranian aggression. In Chapter Seven, we explore potential defensive roles for the Army, including providing training and support to allies and coalition partners and providing reinforcements to augment their defenses. We assess Army ASM concepts, as well as surface-to-surface fires and surface-to-air fires. In addition, we provide some estimates of the cost to field land-based missile systems.

Chapter Eight summarizes our recommendations for U.S. Army investments.

CHAPTER TWO
China in the Western Pacific: Core Interests and Strategic Intentions

Today, the United States and its allies collectively retain the preponderance of economic and military power in Asia. However, trends suggest that China will continue to gain in national power and military capability relative to the United States and its allies.[1] At the same time, long-standing tensions and disputes between China and its neighbors will likely persist. How the United States and its allies should prepare for the confluence of these factors is critical to ensuring the stability of Asian security and the security of the United States and its allies.

In this chapter, we examine China's core interests and strategic intentions, followed by China's disputes with the United States and with Japan, the Philippines (including China's response to the July 2016 South China Sea arbitration ruling in the case of the Philippines versus China),[2] and Taiwan. We end this chapter with a discussion of China's current approach to dealing with such disputes.

---

[1] For an examination of changes that China has made in its military capabilities relative to those of the United States, see Eric Heginbotham, Michael Nixon, Forrest E. Morgan, Jacob L. Heim, Jeff Hagen, Sheng Li, Jeffrey Engstrom, Martin C. Libicki, Paul DeLuca, David A. Shlapak, David R. Frelinger, Burgess Laird, Kyle Brady, and Lyle J. Morris, *The U.S.-China Military Scorecard Forces, Geography, and the Evolving Balance of Power, 1996–2017*, Santa Monica, Calif.: RAND Corporation, RR-392-AF, 2015.

[2] See Permanent Court of Arbitration, "Summary of the Tribunal's Decisions on Its Jurisdiction and on the Merits of the Philippines's Claims," in *The South China Sea Arbitration (The Republic of the Philippines v. The People's Republic of China)*, The Hague, July 12, 2016.

China's Core Interests

China's government has stated that it is pursuing a path of "peaceful development" but that this path will not come at the expense of its "core interests," a term referring to the collective material and spiritual demands of a state and its people that Beijing regards as essential to national development and survival. The 2011 Peaceful Development White Paper outlined the core interests for China, which we capture in Table 2.1.

Table 2.1
China's Core Interests

Core Interest	Description and Threats and Dangers
"State sovereignty"	• Refers to the exercise of authority over all state assets and the rights to dignity and independence from foreign control • Threats include cyber activity that threatens China's sovereign control of its cyber domain, as well as any action by any actor that Beijing deems an affront to the dignity of its people
"National security"	• Refers to the basic security of the nation • Dangers include nuclear war, invasion, separatism, and other threats to the nation's cohesion and survival
"Territorial integrity and national reunification"	• Speaks to the exercise of the Chinese government's authority over all claimed geographic features • Threats include any action by an actor that imperils the integrity of China's borders or the government's control of those features
"China's political system and overall social stability"	• Refers to the Chinese Communist Party–led government and political system • China's leaders regard anything that endangers Chinese Communist Party control and anything that threatens to provoke unrest and instability to be a major threat
"The basic safeguards for ensuring sustainable economic and social development"	• Refers to major sea lines of communication, natural resources, markets, and other economic and financial assets needed for economic growth • Threats include piracy and other threats to Chinese access to these goods

SOURCE: State Council Information Office, *China's Peaceful Development*, white paper, Beijing, September 6, 2011.

China's definition of its core interests underscores the importance that China attaches to Taiwan, the Senkaku (Diaoyu) Islands, and South China Sea maritime territorial disputes. China's leaders view these as crucial because of the implications they hold for political stability, national security, international status, and economic development.

China's Strategic Intentions

China's increase in national and military power is occurring against the backdrop of persistent disputes with several Asian powers. The sources of tension are (1) disputes over sovereignty and territorial claims and (2) disputes over issues of leadership and status in the regional order.[3]

As China's national power has grown, it has sought to reshape the regional and international order to better align with the country's own strategic interests.[4] In addition to issues of leadership and status in the region, tensions have also increased over territorial disputes. China has disputes over its maritime and land borders with many countries, but the most contentious involve those with Taiwan, Japan, Vietnam, the Philippines, and India.[5]

The two sets of security issues are interrelated. China seeks to build a regional and international order that serves its needs of enhancing security, facilitating growth, and elevating its influence to a level commensurate with its status. This requires China to reduce strategic vulnerabilities, especially in the maritime regions, and provides a primary strategic driver for China's pursuit of control over Taiwan and the East and South China Seas. China expresses the strategic value and importance of these national interests through the concept of core

---

[3] Michael P. Colaresi, Karen Rasler, and William R. Thompson, *Strategic Rivalries in World Politics: Position, Space, and Conflict Escalation*, Cambridge, UK: Cambridge University Press, 2008.

[4] Timothy R. Heath, "What Does China Want? Discerning the CHINA's National Strategy," *Asian Security*, Vol. 8, No. 1, 2012.

[5] Office of the Secretary of Defense, *Annual Report to Congress: Military and Security Developments Involving the People's Republic of China 2014*, Washington, D.C.: U.S. Department of Defense, E-6A4286B, April 2014.

interests, as described in Table 2.1. The sections below examine China's perspective on disputes with the United States and key U.S. allies and partners: Japan, the Philippines, and Taiwan.

Disputes with the United States

China does not have any territorial disputes with United States—something that is significant, because territorial disputes have traditionally been regarded as the most likely cause of conflict.[6] China and the United States, however, do have significant disagreements over the nature and direction of the international system. For example, the two countries disagree on the utility and value of U.S. alliances in providing security for Asia. China regards the alliances as generally causing more harm than good, while the United States argues the opposite.[7] Similarly, China and the United States have divergent views about the governance of cyberspace.[8] Many disputes between China and the United States boil down to differences in the status, leadership, and influence of the two countries on a broad array of issues seen as key to their security and development. The most salient of these disputes centers on the question of the political and security order of Asia.

As its power grows, China is seeking to revise or modify elements of the regional security order to better accord with its interests. The recent push by China to promote a "new Asian security concept" reflects this imperative.[9] The United States, in turn, has taken action

---

[6] John Vasquez, *What Do We Know About War?* New York: Rowman and Littlefield Publishers, 2012.

[7] Bates Gill, *Rising Star: China's New Security Diplomacy*, Brookings Institute Press: Washington, D.C., 2010, p. 138.

[8] Kimberly Hsu and Craig Murray, *China and International Law in Cyber Space*, Washington, D.C.: U.S.-China Economic and Security Review Commission, May 6, 2014.

[9] Xi Jinping, "New Asian Security Concept for New Progress in Security Cooperation," remarks at the Fourth Summit of the Conference on Interaction and Confidence Building Measures in Asia, Shanghai, May 21, 2014.

to shore up its influence, status, and leadership in Asia and counter China's challenge, as manifested in the "rebalance to Asia."[10]

As Chinese power grows, the risk and cost of a potential U.S.-China conflict increases. The consequences of war are serious enough that it is in the interest of the United States to find ways to avoid getting drawn into wars that it did not intend. This logic militates against actions and policies that embolden allies into provoking China or that unintentionally encourage escalation and conflict through strategies of military competition.

There are limits, however, to how much the United States can disengage itself from crises and conflicts involving its allies. As China challenges U.S. leadership of the regional order, U.S. credibility becomes increasingly interlinked with the disputes between China and U.S. allies. China's growing rivalry with Japan, a key U.S. ally, adds another layer of strategic significance to the region's territorial and sovereignty disputes. Although it is tempting to dismiss arguments over reefs and rocks as overblown, properly estimating the risks of conflict should acknowledge the strategic dimensions of the various maritime disputes.

U.S. interests in these issues are not insignificant. Ensuring stability and the retention of access and influence in Asia are in the interest of the United States. Allies provide the forward defense of U.S. interests and remain the pillars of an international security order favorable to the exercise of U.S. power. The United States should attempt to balance the need for stability and peace with China with policies that deter aggression and reassure and support U.S. allies.

What is at stake in disputes between China and many of its neighbors goes far beyond the contested features themselves. Because Japan and the Philippines are U.S. allies and because the United States has a long-standing interest in the security of Taiwan, disputes between those nations and China unavoidably carry implications for the relative status and leadership of China and the United States. As competition intensifies, the danger increases that any challenge between China and a U.S. ally will be viewed through a zero-sum lens of increasing or decreasing Chinese and U.S. influence and leadership in Asia. If a crisis

---

[10] Timothy R. Heath, "China and the U.S. Alliance System," *The Diplomat*, June 11, 2014.

escalates to conflict and the United States becomes involved militarily against China, the underlying drive for regional dominance could become difficult to control. This inherent danger is a major reason why conflict between major powers—especially those of a system leader, such as the United States, and an aspiring regional leader, such as China—has traditionally been associated with the most destructive types of wars.[11]

Thus, the United States faces a considerable challenge in making credible its support to its allies in defense of their interests against potential Chinese coercion and aggression. With U.S.-China relations defined by economic interdependence and overall cooperation, the United States remains reluctant to engage in the sort of hostile competition that characterized its relationship with the Soviet Union in the Cold War. Because the risks of confrontation and conflict are so high, the United States has increasingly looked to its allies to take on a growing share of the burden of deterrence. But this approach has inherent limits because of the underlying competition between China and the United States.

On the Chinese side, China attaches considerable importance to maintaining a stable relationship with the United States. For China, a stable relationship is essential for regional and global stability, which it needs for national development. Avoiding a costly and debilitating confrontation with the United States appears to be one of the central goals of the "new type of great power relationship" that Beijing has proposed for U.S.-China ties.[12] At the same time, however, China sees the United States as the greatest potential threat to its core national security interests and objectives, a perspective that is informed not only by its perception that the United States is determined to prevent China from challenging the position of the United States as the world's leader but also by its interpretation of a number of specific incidents, such as

---

[11] David Rapkin and William Thompson, *Transition Scenarios: China and the United States in the Twenty-First Century*, Chicago: University of Chicago Press, 2013.

[12] Michael S. Chase, "China's Search for a 'New Type Great Power Relationship,'" *China Brief*, Vol. 12, No. 17, September 7, 2012.

the accidental U.S. bombing of the Chinese Embassy in Belgrade in May 1999.

Chinese leaders also regard the U.S. alliance system in Asia as a threat. President Xi Jinping has declared that it is "disadvantageous" for Asia if countries "strengthen military alliances with third parties."[13] Given this declaration, the U.S. policy of "rebalancing" to Asia is clearly a source of concern. Many Chinese observers see the rebalancing policy as aimed at consolidating and strengthening the United States' regional alliances and enhancing U.S. military power in the western Pacific. Many Chinese national security analysts also view the U.S. military's development of ASB as aimed squarely at China.[14]

Chinese scholars and analysts debate all these issues, and some raise doubts about the willingness and the ability of the United States to sustain its focus on Asia given its contentious partisan politics, budget difficulties, and competing priorities in other parts of the world. However, there appears to be a relatively broad consensus that the United States poses a serious potential threat to what China views as its most important security interests, even as Chinese officials acknowledge that a cooperative relationship with the United States also remains critical for enabling the country's rise.

Sovereignty Disputes with Japan

Japan has both territorial disputes and status and leadership issues with China. In terms of territory, China and Japan disagree on the ownership of the Senkaku Islands, located in the Ryukyu Island chain. Japan controls the islands and refuses to acknowledge that the islands are in dispute. In 2012, Japan's national government purchased the islands

---

[13] *Xinhua*, "Xi Jinping's Remarks at the Fourth Conference on Interaction and Confidence Building Measures," May 21, 2014a.

[14] Furthermore, Li Yan stated, "For the Americans have said very clearly that AirSea Battle is mainly directed at anti-access and area denial warfare, and [past U.S. assessments] all show that they believe China is conducting anti-access and area denial warfare." Quoted in Kathrin Hille, "U.S. Seeks to Calm Beijing Containment Fears," *Financial Times*, December 8, 2011.

from an individual who held nominal private control.[15] Although official Chinese documents do not refer to the Senkaku Islands disputes in the same way as they do Taiwan, these issues are linked to China's core interests in the sense that Beijing sees the islands as involving Chinese sovereignty and territorial integrity, economic development opportunities, and domestic stability issues.

The Senkaku Islands consist of roughly eight islands and an array of lesser islands and reefs. The total landmass area of the islands measures about 5.69 square kilometers. The largest island—Uotsuri Shima (Diaoyu Dao in Chinese)—has a landmass of about four square kilometers, and its highest elevation reaches 383 meters. The second largest, located 27 kilometers to the northeast, measures roughly one square kilometer in area. Each of the other islands measures less than half a square kilometer in area.[16] The islands are roughly 300 kilometers from China's coast and 400 kilometers from Okinawa.[17] The Senkaku Islands are economically important to both countries because the islands sit near potentially lucrative natural gas fields. In addition, these islands have strategic value because they lie astride Japan's sea lines of communication and to the north of Taiwan. Control of the islands could give either country influence in any Taiwan-related crisis. The islands also affect China's ability to project power further into the Pacific Ocean.

Prior to the 1970s, China did not make any official statements claiming the Senkaku Islands. In fact, a number of Chinese maps leading up to the 1970s appear to acknowledge Japanese ownership of the islands, spurring Chinese authorities in subsequent years to crack down on "erroneous maps."[18] China's position changed after surveys in 1968 uncovered the possibility of oil reserves near the islands and after the United States ended its administration of the Ryukyu Islands in 1971.

---

[15] Julian Ryall, "Japan Agrees to Buy Senkaku Islands," *Telegraph*, September 5, 2012.

[16] State Council Information Office, *Diaoyu Dao, an Inherent Territory of China*, Beijing, September 24, 2012.

[17] Ministry of Foreign Affairs of Japan, "Japanese Territory: Senkaku Islands," web page, April 13, 2014a.

[18] *Xinhua*, "China Cracks Down on Erroneous Maps," January 9, 2013a.

China maintains that when the two countries were normalizing relations in the early 1970s, they reached an understanding on "leaving the issue . . . to be resolved later" and contends that recent Japanese actions, most notably the 2012 nationalization of several of the islands, have undermined this consensus.[19] In addition, Beijing also strongly objects to Tokyo's position that it will not acknowledge the existence of a dispute over the islands.[20]

In its official white paper on the islands, issued in 2012, Beijing insisted that the islands are part of China's "inherent territory." The same document declared that China's "will to defend national sovereignty and territorial integrity is firm."[21] Officials now routinely state that China has "indisputable sovereignty" over the islands.[22]

To support its claims to the islands, China employs historical, legal, media, and administrative measures.[23] China has used arguments drawing from historical maps, travelogues, and other documents to suggest long-standing control of the islands. It has also begun to pass laws and issue official documents claiming ownership of the islands. In 1992, China passed the Law on the Territorial Sea and Contiguous Zone, which declared that the islands "belong to China." The 2009 Law on the Protection of Offshore Islands outlined management and administrative measures for control of all islands. The Chinese government issued a statement on September 10, 2012, announcing territorial baselines for the islands. China has also increased its maritime patrol

---

[19] Japan maintains that there has never been any agreement to shelve the issue of sovereignty over the islands. See Ministry of Foreign Affairs of Japan, "Japanese Territory: Senkaku Islands—Situation of the Senkaku Islands," web page, April 4, 2014a.

[20] The website of Japan's Ministry of Foreign Affairs highlights Tokyo's position that "[t]here exists no issue of territorial sovereignty to be resolved concerning the Senkaku Islands." See Ministry of Foreign Affairs of Japan, "Japanese Territory: Senkaku Islands," web page, April 13, 2014b.

[21] State Council Information Office, 2012.

[22] *Xinhua*, "FM Spokesman Comments on Japan's Statement of Defense on Senkaku," November 22, 2014b.

[23] International Crisis Group, "Dangerous Waters: China-Japan Relations on the Rocks," *Asia Report*, No. 245, April 8, 2013.

presence significantly since 2009.[24] In 2013, China declared an Air Defense Identification Zone that encompassed the islands.[25] Occasionally, Chinese protestors have landed on the islands, but these individuals have been quickly seized and removed by Japanese authorities.[26]

Chinese officials also increasingly pledge retaliation for any action that challenges Chinese sovereignty over its maritime claims. The 2013 Defense White Paper repurposed Mao Zedong's dictum—"We will not attack unless we are attacked; but we will surely counter attack if attacked"—from one of defense against invasion and nuclear attack to one that warns neighboring powers that China will "resolutely take all measures necessary to safeguard its national sovereignty and territorial integrity."[27] China has used a strategy of reacting to perceived provocations by Japan by taking strong countermeasures to change the status quo in its favor.[28]

In terms of status and leadership issues, the territorial dispute is aggravated by intense disputes related to the status and influence of each country in the Asia-Pacific region. Long regarded as the leading economic power in Asia, Japan has responded to its eclipse by increasing its diplomatic, security, and economic outreach to countries nervous about China's growing power. Japanese Prime Minister Shinzo Abe has worked hard to deepen bilateral and multilateral engagements throughout South and Southeast Asia. Japan has also provided maritime security platforms to Vietnam and the Philippines, two countries that share with Japan intense sovereignty disputes with China. Japan has also countered Chinese economic initiatives with its own proposals, as Abe did when visiting India in early 2014. These activities raise

---

[24] State Council Information Office, 2012.

[25] *BBC News*, "China Establishes Air Defense Zone over East China Sea," November 23, 2013.

[26] *Japan Times*, "Chinese Activists Land on Senkaku Islet; Japan Arrests 14," August 16, 2012.

[27] Quoted in *Xinhua*, "The Diversified Employment of China's Armed Forces," April 19, 2013c. See also, State Council Information Office, *The Diversified Employment of China's Armed Forces*, white paper, Beijing, April 2013.

[28] International Crisis Group, 2013.

the possibility that any conflict between China and Japan, or between China and a country with which Japan is enhancing security ties, could rapidly multilateralize, posing extra challenges to deescalation.

Sovereignty Disputes with the Philippines

The Spratly Islands are a collection of islands and reefs in the middle of the South China Sea. These are mostly uninhabited features with little arable land, and only a few islands carry any fresh water. The total landmass of all features in the Spratly Islands measures roughly four kilometers, not including any of the area added by Chinese dredging and construction operations. The Spratly Islands, however, feature fertile fishing grounds, as well as potentially lucrative gas and oil reserves. The islands are also located astride key shipping lanes through which much of Asia's trade currently passes.

The Republic of China, which preceded the People's Republic of China, first declared ownership over the Spratly Islands in the 1930s and 1940s in response to French assertions of sovereignty. After the establishment of the People's Republic of China in 1949, Beijing continued to uphold the "nine-dashed line" claim that encompasses all of the Spratly Islands.[29]

China's official position is that it has "indisputable sovereignty" over all the "land features and adjacent waters" in the South China Sea. However, China has offered to negotiate "maritime rights" and "territorial sovereignty" issues bilaterally with rival claimants. China has also proposed to shelve the question of ownership in favor of "joint development" of resources. At the same time, China has used historical, legal, media, administrative, and military means to consolidate its de facto control of the islands.[30] It has clashed militarily with Vietnam

---

[29] International Crisis Group, "Stirring Up the South China Sea," *Asia Report*, No. 223, April 23, 2012.

[30] *Xinhua*, "Full Text: CHINA Government Position Paper on Matter of Jurisdiction in South China Sea Arbitration Initiated by Philippines," December 7, 2014c.

in the 1970s and 1980s over control of the Paracel Islands, which lie north of the Spratlys.[31]

China has also seized control of several features claimed by the Philippines in the Spratly Islands. In 1995, China occupied Mischief Reef, building a structure believed to house roughly 40 People's Liberation Army (PLA) marines. China occupies nine reefs and continues to upgrade and reclaim land in at least five of them—South Johnson, Cauteron, Hughes, Gaven, and Eldad Reefs. Philippine media reported that 200 PLA troops occupied Fiery Cross and Subi Reefs in 2014.[32] China also exercises de facto control of many other nearby features.

In 2012, China seized control of Scarborough Reef from the Philippines through tactics widely described as "salami slicing" in Western press and as a "cabbage strategy" in the Chinese press.[33] China relied on superior numbers of civil maritime security forces edging out the Philippine counterparts, combined with media and legal campaigns and threats of economic retaliation to pressure the Philippines against escalating the situation.

Sustaining a military presence on the Spratly Islands poses a considerable logistics challenge for the PLA. Most of the Spratly features are more than 600 nautical miles from Hainan Island, and the arid, tiny features render personnel and equipment posted there almost completely dependent on shipborne supplies.

China's leaders have hardened their stance about territorial and sovereignty claims in the South China Sea in recent years, as they did with the Senkaku Islands. Chinese President Xi Jinping has reiterated that China will "absolutely not give up [its] legitimate rights and interests, and will definitely not sacrifice the state's core interests." He warned, "Foreign countries should not expect that we will trade our own core interests" and "not expect that we will eat the bitter fruits

---

[31] Thomas J. Cutler, *The Battle for the Paracel Islands*, Annapolis, Md.: Naval Institute Press, 1974.

[32] D. J. Sta. Ana, "China Reclaiming Land in Five Reefs?" *Philippine Star*, June 13, 2014.

[33] See, for example, Harry Kazianis, "China's Expanding Cabbage Strategy: After Pursuing a Cabbage Strategy in the South China Sea for Years, Could Beijing Adopt This Against Japan?" *The Diplomat*, October 29, 2013.

of damaging our country's sovereignty, security, and developmental interests."[34]

In July 2016, the Permanent Court of Arbitration issued its award in the case the Philippines brought against China with respect to its claims in the South China Sea and its behavior in disputes over the Spratly Islands and Scarborough Shoal.[35] Most experts anticipated that the tribunal's award would largely favor the Philippines, but it exceeded expectations by delivering a sweeping rebuke of Chinese assertions and actions. China has consistently stated that it would ignore the award, although international legal scholars almost uniformly consider it to be binding on both parties, despite the fact that China refused to participate in the proceedings.

The award addressed the following major points, which are described in more detail below:

- China's claim of "historic rights" and the nine-dash line
- status of features in the South China Sea and their maritime entitlements
- lawfulness of Chinese actions
- harm to the marine environment
- aggravation of the dispute by China.

As to China's *claims of historic rights* in the South China Sea and its representations on the *nine-dash line*, the tribunal found in favor of the Philippines. The tribunal concluded: "[T]o the extent China had historic rights to resources in the waters of the South China Sea, such rights were extinguished to the extent they were incompatible with the exclusive economic zones provided for in the Convention." Additionally, the tribunal found there was "no legal basis for China to claim historic rights to resources within the sea areas falling within the 'nine-dash line.'"[36]

---

[34] *Xinhua*, "Xi Jinping Stresses at the Third Collective Study Session of the Political Bureau to Make Overall Planning for Domestic and International Situations," January 29, 2013b.

[35] For a summary of the award, see Permanent Court of Arbitration, 2016.

[36] Permanent Court of Arbitration, 2016.

On the question of the *status of features in the South China Sea*, the tribunal award also found in favor of the Philippines, ruling that none of the features claimed by China met the standard of an island and thus none was entitled to a 200 nautical mile exclusive economic zone (EEZ). The award found instead that the features in question include rocks that are entitled to a 12–nautical mile territorial sea, as well as low-tide elevations that generate no maritime entitlements. Consequently, the tribunal concluded: "[N]one of the Spratly Islands is capable of generating extended maritime zones" beyond 12 nautical miles, and "the Spratly Islands cannot generate maritime zones collectively as a unit." Additionally, the tribunal stated: "Having found that none of the features claimed by China was capable of generating an [EEZ], the Tribunal found that it could—without delimiting a boundary—declare that certain sea areas are within the [EEZ] of the Philippines, because those areas are not overlapped by any possible entitlement of China."[37] This finding had important implications for the question of the lawfulness of China's actions.

On the question of *whether China's actions were lawful*, the tribunal stated that, because certain areas are within the EEZ of the Philippines, "China had violated the Philippines' sovereign rights in its [EEZ] by (a) interfering with Philippine fishing and petroleum exploration, (b) constructing artificial islands and (c) failing to prevent Chinese fishermen from fishing in the zone." The tribunal also found that fishermen from the Philippines and from China have traditional fishing rights at Scarborough Shoal and that China "had interfered with these rights in restricting access." In addition, the tribunal further held that Chinese maritime law enforcement vessels "had unlawfully created a serious risk of collision when they physically obstructed Philippine vessels."[38]

As to the issue of *harm to the marine environment*, the tribunal found that China's massive land reclamation and island-building "had caused severe harm to the coral reef environment and violated [China's] obligation to preserve and protect fragile ecosystems and the habitat of

---

[37] Permanent Court of Arbitration, 2016.

[38] Permanent Court of Arbitration, 2016.

depleted, threatened, or endangered species." The tribunal also found that China had been well aware that Chinese fishermen "have harvested endangered sea turtles, coral, and giant clams on a substantial scale in the South China Sea (using methods that inflict severe damage on the coral reef environment)" and that China had failed to fulfill its obligations to put a stop to such activities by its fishers.[39]

In the final area, the issue of whether China's actions since the commencement of the arbitration case constitute *aggravation of the dispute*, the tribunal found that it did not have jurisdiction to consider actions by Chinese military and law enforcement vessels at Second Thomas Shoal, but that China's

> large-scale land reclamation and construction of artificial islands was incompatible with the obligations on a State during dispute resolution proceedings, insofar as China has inflicted irreparable harm to the marine environment, built a large artificial island in the Philippines' exclusive economic zone, and destroyed evidence of the natural condition of features in the South China Sea that formed part of the Parties' dispute.[40]

On the whole, the award was a stunning and almost total rebuke of China's actions and statements about its dispute with the Philippines in the South China Sea. Beijing swiftly repudiated the decision and reiterated its arguments that the proceedings were biased and invalid. China also continued to exaggerate support for its position from other countries, as it had done for some time prior to the announcement of the tribunal's award. China also warned the Philippines to pursue bilateral negotiations on Chinese terms. As one official Chinese media article put it: "The choice between a path of confrontation, conflict and economic risks and a road of friendship, peace and prosperity is a simple one."[41] Ahead of the ruling, Beijing also criticized the United States for what it called a double standard in demanding that China

---

[39] Permanent Court of Arbitration, 2016.

[40] Permanent Court of Arbitration, 2016.

[41] *Xinhua*, "Commentary: Philippines Wise to Draw Experience from U.S. Bloody Record of Intervention," July 10, 2016.

adhere to the ruling even though the United States refused to accept a ruling from the International Court of Justice in a case involving Nicaragua in the 1980s.[42] In addition to its diplomatic response, China also signaled its displeasure with military activities conducted just prior to and shortly after the July 12, 2016, announcement of the award.

Just ahead of the award, the PLA Navy (PLAN) conducted an exercise in early July 2016 covering "air and sea areas from Hainan Island to the Xisha Islands." The Chinese Ministry of National Defense (MND) stated that participating forces included "naval vessels, fixed wing aircraft, helicopters, and other equipment."[43] Chinese media reports added that the exercise involved guided missile destroyers and frigates, including ships from all three PLAN fleets. Training topics reportedly included anti-air warfare, anti-submarine warfare, and anti-surface warfare, and pictures carried by official media showcased a PLAN frigate launching an anti-ship cruise missile and frigates and destroyers launching SAMs. The MND asserted that the exercise was "a routine arrangement made in accordance with the annual training plan of the Chinese Navy."[44] The *Global Times*, a Chinese newspaper that is state-owned but much more provocative and freewheeling than more-authoritative official media outlets, reported that the exercise was not linked to the arbitration ruling, an assertion that most observers outside China found difficult to believe given the timing.[45] In the region and in the United States, many analysts concluded that the timing was clearly linked to China's broader diplomatic response to the arbitration case. Indeed, even if the exercise dates were set before the announcement about when the ruling would be forthcoming, China

---

[42] Quan Xianlian, "The Right to Reject Tribunal Ruling Is Real," *China Daily*, July 11, 2014.

[43] *China Military Online*, "Chinese Navy to Conduct Drills Around Xisha Islands," July 6, 2016.

[44] Yao Jianing, "Chinese Navy to Conduct Drills Around Xisha Islands," *China Military Online*, July 6, 2016.

[45] Shan Jie, "China to Hold Drills Near Xisha Islands: Exercises in S. China Sea Not Linked to Arbitration: Expert," *Global Times*, July 4, 2016.

still could have rescheduled the exercise if it wanted to avoid creating the impression that it was part of the reaction to the ruling.

In addition, along with other Chinese military activities, some observers viewed the naval exercise as a response to U.S. military operations in and around the South China Sea. According to a *Global Times* editorial: "If the US is taking advantage of the mess to deploy more military forces to the South China Sea, which are a direct threat to China's national security, China's military exercises could be regarded as a countermeasure."[46] Somewhat similarly, an MND spokesperson stated:

> [The] frequent activities of the US ships and planes in the South China Sea . . . are meant to make a show of force and to weaken the determination and will of China to protect national sovereignty and security. It is an act of militarization in the South China Sea and it endangers regional peace and stability. But I'd like to say that the US side is making the wrong calculation. The Chinese armed forces never give in to outside forces. If it is our territory, we will definitely safeguard it. If it is not ours, we will not ask for it. The will and capability of China to safeguard national sovereignty and territorial integrity are very firm.[47]

Following the announcement of the award, China underscored the point about the exercises being a countermeasure by conducting air operations in the vicinity of Scarborough Shoal and publicly releasing images of a PLA Air Force (PLAAF) H-6K bomber flying over the area with the disputed feature clearly visible in the background.[48] Along with the pictures the PLAAF issued a statement indicating that such patrols involving H-6K bombers, fighter jets, and reconnaissance planes would continue as needed for the purposes of improving training and bolstering China's sovereignty claims. "Based on the need of

---

[46] *Global Times*, "Power Game Decides Post-Arbitration Order," July 5, 2016.

[47] Ministry of National Defense of the People's Republic of China, "Defense Ministry's Regular Press Conference on June 30," June 30, 2016.

[48] Anders Corr, "Chinese Bomber Buzzes Philippines' Scarborough Shoal in Latest Salvo of U.S.-China Signaling War," *Forbes*, July 17, 2016.

the Air Force for fulfilling its missions and tasks, the combat readiness patrol to the South China Sea by the Air Force servicemen will continue on [a] regularized basis," the PLAAF spokesperson said.[49]

Sovereignty Disputes with Taiwan

Taiwan has long been viewed by China as the most central issue for its national security. Then–State Councilor Dai Bingguo stated in a 2010 speech: "The Taiwan question constitutes China's core interest concerning its unification and territorial integrity, dear to the heart of the 1.3 billion Chinese citizens and the whole Chinese nation."[50] Beijing's preferred approach is "peaceful unification," and the remarkable progress in the cross-Strait relationship since 2008 has made the Taiwan Strait much less of a flash point than it has previously been. Nonetheless, China continues to strengthen its ability to use force against Taiwan. Taiwan's MND has stated that China's goal is to ensure that the PLA will be able to use force against Taiwan by 2020, if called on to do so by Chinese Communist Party leaders.[51] Even if China does not use force, improving its ability to do so strengthens its bargaining leverage and threatens to erode Taiwan's ability to protect its interests as it engages with the mainland.

The legal basis for Beijing's position on unification with Taiwan is China's "anti-secession law" passed in 2005 during a much more turbulent period in cross-Strait relations. According to the law, China would use force under the following conditions:

> In the event that "Taiwan independence" secessionist forces should act under any name or by any means to cause the fact of Taiwan's secession from China, or that major incidents entailing

---

[49] Zhang Yunbi, "China's Air Force Flags Regular Patrols in South China Sea," *China Daily*, July 19, 2016.

[50] Dai Bingguo, "Adhere to the Path of Peaceful Development," Ministry of Foreign Affairs, People's Republic of China, December 6, 2010.

[51] Ministry of National Defense of the People's Republic of China, *2013 Republic of China National Defense Report*, Beijing, October 8, 2013.

Taiwan's secession from China should occur, or that possibilities for a peaceful unification should be completely exhausted, the state shall employ non-peaceful means and other necessary measures to protect China's sovereignty and territorial integrity.[52]

Similarly, a speech by Dai stated that the Chinese "will never allow Taiwan to split from China, nor will [they] ever commit [themselves] to the renunciation of force."[53]

Current Chinese Focus on Resolving Sovereignty Disputes

Given China's territorial ambitions and its aspirations for greater regional influence, the highest potential for crisis and conflict lies in disputes between China and its neighbors that escalate into armed conflict. Because of U.S. security commitments, a conflict involving Beijing and a U.S. ally would almost unavoidably involve Washington at some level.

China's behavior to date has tended toward an aversion to high-risk military attacks to seize territory. Instead, China has found it more promising to use diverse economic, diplomatic, and political levers to pressure recalcitrant countries into accommodating Chinese preferences. This "salami slicing" strategy has worked well in expanding Chinese control of its core interests in a manner that minimizes the risk of instability.

While a peaceful resolution of China's disputes with its neighbors should be encouraged, conflicts that disrupt the region's growth and prosperity cannot be ruled out. A China frustrated by a lack of progress on many policy issues could be tempted to become increasingly assertive in a way that raises the risk of a military clash. An intensifying rivalry between China and the United States, or between China and

---

[52] 10th National People's Congress and Chinese People's Political Consultative Conference, 3rd session, Anti-Secession Law, Beijing, March 14, 2005.

[53] Dai, 2010.

Japan, would also greatly increase the overall risk of conflict, especially if Beijing began to conclude that time was no longer on its side.

The next three chapters examine the situation from the standpoints of Japan, the Philippines, and Taiwan, respectively, and discuss some scenarios about how conflict could emerge between China and each of those three countries and what this might mean for the United States.

CHAPTER THREE

China-Japan Relationship from Japan's Standpoint

In this chapter, we examine the China-Japan relationship from the Japanese standpoint. We start by providing some context on the Japan-China relationship and then examine some potential conflict scenarios between the two countries, Japan's evolving security posture and operational concept, and how the U.S. Army can help Japan help itself.

Context: U.S. Relationship with Japan

The U.S.-Japan security relationship has grown in importance to both countries in recent years. The strength of the relationship rests on the history of U.S. military presence in Japan since 1945 and on subsequent mutual defense treaties, but the post–World War II design of the relationship and Japan's constrained view of its national security role have limited the security alliance in many ways when compared with U.S. alliances with NATO and South Korea.

The United States and Japan continue to uphold a defense treaty signed in 1960 (Treaty of Mutual Cooperation and Security Between the United States and Japan), wherein each party has pledged that an "armed attack" against "either party in the territories under the administration of Japan" would be regarded as "dangerous to its own peace and safety." In such an event, the treaty obligates each party to "meet the common danger in accordance with its constitutional provisions and processes."¹ Senior U.S. officials, including the Presidents Barack Obama and Donald Trump, have stated that the United States regards the treaty as applicable to the Senkaku Islands, because these remain under the "administrative control of Japan."²

Potential China-Japan Conflict Scenarios

Although uninhabited, the Senkaku Islands are claimed by both China and Japan. Japan has administered the Senkakus since annexing them in 1895, except for a period of U.S. occupation control from 1945 to 1972. The Japanese briefly settled the islands, with a small settlement operating a fish-processing facility from 1900 to 1940. Japan has declared its intent to maintain control of the Senkakus and has positioned forces in the region to defend its southwest Ryukyu Islands chain.

The Chinese contest Japanese sovereignty over the Senkakus. China claims ownership of the islands—which it calls the Diaoyus—dating back to the 15th century. China frequently sends ships and airplanes to the Senkakus and is pressing its claim through diplomatic and media channels. Japan claims that, in 2015, vessels of the Chinese government entered Japanese territorial waters on at least 35 occasions. Japan also claimed that, in 2015, China sent an armed coast guard vessel into Japanese territorial waters for the first time.³ These incursions have mostly remained at the nuisance level, and Japan has shown a high level of competence in monitoring them and sending forces to protect its territorial waters.

For its part, the United States recognizes neither nation's claims but does recognize Japanese administrative control over the Senkakus.

---

[1] U.S. Department of State, "U.S. Collective Security Agreements," web page, undated.

[2] *Washington Post*, "Q&A: Japan's Yomiuri Shimbun Interviews President Obama," April 23, 2014; White House, Office of the Press Secretary, "Remarks by President Trump and Prime Minister Abe of Japan in Joint Press Conference," February 10, 2017.

[3] James Mayger and Yuji Nakamura, "Japan Protests Intrusion of Armed Chinese Vessels into Its Waters," *Bloomberg*, December 25, 2015.

The United States has promised to come to Japan's defense if its forces are attacked. President Barack Obama stated that any attack on Japanese forces, including in the Senkakus, invokes Article 5 of the Japan-U.S. mutual defense treaty.[4]

The Senkakus provide the most likely flash point for a conflict between China and Japan. The contest over the islands is aggravated by a growing competition for influence and leadership in Asia. Several studies have established that the risk of a military clash is highest when it occurs within the context of strategic rivalry and following a series of crises over a particular issue, usually involving disputed territory.[5] Thus, China and Japan face a heightened risk of a clash if the two countries experience repeated major crises near the disputed islands. If the strategic rivalry between China and Japan intensifies, there is a risk that such confrontations could escalate into an armed conflict.

Here we examine two categories of conflict scenarios: (1) scenarios short of military seizure of the islands and (2) military seizure of the islands.

Scenarios Short of Military Seizure of Islands

China may be expected to continue pursuing coercive behavior by deploying fishing vessels; coast guard ships; and strong political, media, and diplomatic pressure against Japan below the threshold of military conflict. Because at this point neither side seeks a war, the odds of escalation remain low. Tensions over the Senkakus could become a crisis if activists from either country are arrested for landing on the islands. Such a crisis could also emerge out of a collision of fishing or maritime law enforcement vessels or even from an accident involving military platforms. Because of the relative parity of power between China and Japan, it would be difficult for either side to consistently prevail in such incidents. As such, a stalemate—a deepening of frustration and

---

[4] White House, "Remarks by President Obama and Prime Minister Abe of Japan in Joint Press Conference, Rose Garden," Washington, D.C.: Office of the Press Secretary, April 28, 2015.

[5] Brandon Prins, "Interstate Rivalry and the Recurrence of Crises: A Comparison of Rival and Nonrival Crisis Behavior, 1918–1994," *Armed Forces and Society*, April 1, 2005. See also Colaresi, Thompson, and Rasler, 2008.

hostility and an intensification of strategic rivalry—is the most likely outcome. This dynamic raises the likelihood of another crisis, which would in turn contribute to a worsening of the rivalry dynamic and overall elevation of the risks of conflict.

In an atmosphere of intense acrimony, rivalry, and distrust following a series of militarized crises, any incident could rapidly escalate to a clash of arms. This could start as an exchange of fire between coast guard ships, but the situation could quickly escalate if military platforms were nearby. Although a tense standoff that is eventually defused through diplomatic action is the most likely scenario, an exchange of fire causing injury or the loss of life or damage or destruction to platforms cannot be ruled out. With emotions running high and temperatures flaring, leaders would have to calculate the risks of escalation. The danger of escalation into a major war between China and Japan should put intense pressure on Chinese, Japanese, and U.S. leaders to find a way to stave off or deescalate any such situation.

Because of the political opprobrium of aggression and the risk of U.S. involvement, an unprovoked Chinese assault on the Senkakus would offer little benefit to China and carry extremely high risks. A more plausible scenario would be a spiraling of intensifying and protracted crises with little resolution and a deepening of suspicion and hostility that resulted in a clash in which China sustained losses of personnel. This would provide a powerful incentive for China to seek some sort of reciprocal military action if Beijing could plausibly characterize Japanese actions as aggressive. It would then have a pretext to strike Japanese forces in a punitive operation with limited objectives that could be achieved in a short time.

Examples could include missile strikes to sink a Japanese naval combatant or down a Japanese fighter aircraft that flew too close to Chinese forces. The downside to this course of action is that Japan could in turn seek to escalate the conflict, raising the risk that the two countries could head toward war.

Military Seizure of Islands Scenarios

A clearly aggressive Chinese act to seize control of the Senkakus, whatever the pretext, would constitute a high-risk action that would almost certainly compel a U.S. response. The principal risks for China would be the difficulty of sustaining forces deployed on the island and the vulnerability to counterattack by Japanese and U.S. forces. Should Japan decide to contest the seizure or to evict the Chinese presence through military or legal means, conflict would be difficult to avoid and escalation just as hard to control. And if the United States decided to join Japan's effort to halt or reverse the seizure, conflict could escalate rapidly. Should China find its forces evicted from the Senkakus, the humiliation of a major defeat could seriously destabilize the Chinese government.

Another possible scenario envisions the placement of military or law enforcement personnel on the island to create a fait accompli seizure. If the law enforcement personnel were uniformed and armed, this would amount to a variation on the military attack scenario, bearing the same vulnerabilities and drawbacks. Another variation of this scenario could involve government personnel disguised as civilians who secretly occupy the island. The problem with this stratagem is that its success depends on the publication of the fact that China has gained control of the islands through military or civil authorities. If the individuals only carried light arms in a bid to maintain the disguise, then this would leave the Chinese personnel extremely vulnerable. Thus, this course of action would carry all the risks of the other island seizure scenarios but offer the least protection of the deployed forces.

A last variation of the seizure scenario considers the secret placement of individuals from nongovernment organizations. While this may seem appealing as a way to challenge Japanese control in a manner that undercuts the rationale for a U.S.-Japan military response, it also loses the potency of message that comes with actual Chinese government control of the islands. In fact, the presence of nonofficial activists has occurred numerous times before. In all cases, Japanese authorities simply removed the activists before problems of sustainment could threaten their lives.[6] A repeat of this tactic would likely gain China very little.

---

[6] *Japan Times*, 2012.

Addressing the China Threat from Japan's Standpoint

Japan's Security and Military Posture in Response to the China Threat

Japan is undergoing a fairly substantial change to its military: moving slowly away from its Cold War outlook and posture that focused exclusively on threats to Japan and that were oriented north. Japan is slowly shedding the most-insular aspects of its strategic concept while moving toward a new force focused more on protecting its offshore islands. With the election of Prime Minister Abe in 2012, Japan's security policy received its most critical review, and his administration has pushed for a substantial shift in that policy. The security constraints in the Japanese constitution have been reinterpreted to allow a much more forward-leaning security policy. This should allow Japanese forces to contribute more directly to collective security activities, but the specifics are still emerging. The United States and Japan released the revised Guidelines for Japan-U.S. Defense Cooperation in April 2015.[7] The guidelines signal that both sides intend to develop closer bilateral security ties. The guidelines also open the possibility of greater military support to the United States in conflicts involving another U.S. ally. Finally, they addresses greater cooperation on regional security issues, including building partner capacity, maritime security, and logistics support. Japan has relaxed the arms-export ban and is now in discussions with several countries to take advantage of the changes.

Although Japan's strategic outlook is evolving significantly under the Abe administration, the starting point was a very narrow conception of security; therefore, despite significant movement, Japan's security outlook is likely to remain more constrained and inward-looking, compared with the outlooks of other major powers. For example, Japan's latest National Defense Program Guidelines promises a more "proactive contribution to peace," but the document also reaffirms that Japan's investments in its defense force are being made while "maintaining an exclusively defense-oriented policy, not becoming a military

---

[7] Ministry of Defense of Japan, *The Guidelines for Japan-U.S. Defense Cooperation*, Tokyo, April 27, 2015a.

power that poses a threat to other countries."[8] This is also reflected in the reinterpretation of the Japanese constitution, a reinterpretation that was adopted in 2014 with the condition that the expanded scope applies only under an attack that "threatens Japan's survival" and that Japan must limit force to the "minimum extent necessary."[9] Still, while Japan continues to have major limitations on its military forces, including budgetary, it has been able to field some significant military capabilities.

Japan's Current Operational Concept in Response to the China Threat

Japan is looking to make substantial changes to its self-defense force to shift it from its previous orientation toward the north and Russia to a force oriented toward the south, where the Senkaku Islands lie, and to creating capabilities to protect those interests more dynamically: seeking air and maritime superiority and capabilities to rapidly deploy ground forces. Japan's military spending remains at 1 percent of gross domestic product (GDP), and, in 2016, its defense budget was about $40 billion (4.861 trillion yen).[10]

Japan is most concerned about potential threats from China and North Korea. In recent years, China has made the Senkaku Islands a bilateral issue that has involved a range of tests, including military, for Japan. North Korea has threatened Japan with ballistic missile attacks, and North Korean agents have kidnapped Japanese citizens.

To address these and other challenges, Japan's Self-Defense Forces (SDF) has 11 primary missions:[11]

1. ensure security of sea and airspace surrounding Japan

---

[8] Ministry of Defense of Japan, *National Defense Program Guidelines for FY 2014 and Beyond*, Tokyo, December 17, 2013b, p. 6.

[9] Ministry of Foreign Affairs of Japan, "Cabinet Decision on Development of Seamless Security Legislation to Ensure Japan's Survival and Protect Its People," web page, July 1, 2014c.

[10] Ministry of Defense of Japan, *Defense of Japan 2016*, Tokyo, December 2015b, p. 41.

[11] Ministry of Defense of Japan, 2015b, p. 283.

2. defend Japan's remote islands
3. respond to ballistic missile attack
4. respond to attacks by guerillas, special operations forces, and others
5. initiatives toward ensuring maritime security
6. responses in space
7. response to cyber attacks
8. response to large-scale disasters
9. transport of Japanese nationals overseas
10. readiness against invasion
11. response to other events.

Japan's proposed future investments reflect these priorities. Japan operates a number of air and maritime sensor platforms, as well as ground-based radars, to monitor its air and maritime sovereignty. In doing so, Japan has documented a growing number of PLA violations of its sovereignty. Some of these capabilities are optimized to monitor small craft of the type that North Korea might use in the kidnapping of Japanese citizens.

Defending against attacks on remote islands has surged to the fore of Japanese security thinking in recent years and is arguably forcing the largest changes in the SDF. While this approach is likely influencing investments in the air and maritime force for new fighter aircraft, early warning aircraft, Aegis destroyers, and submarines, it is also having an impact on the Ground SDF (GSDF). Half the GSDF is being reorganized and equipped to become rapid deployment units with an orientation toward the south, where the Senkakus lie. In part, the mobility to implement this concept will come from investing in more than 50 amphibious landing craft (AAV7s) and tilt-rotor aircraft (V-22s), both of which could be used to quickly place ground forces on offshore islands. In addition, these forces could be deployed using two existing Hyuga-class helicopter-landing ships and soon by using Japan's two new Izumo-class helicopter carriers. Japan's current medium-term defense program announced plans to refurbish its three existing Osumi-class landing ship, tanks (LSTs), and to consider whether it should procure a large amphibious ship capable of launching fixed-wing aircraft.[12] Beyond the technology and organizational changes, the GSDF has been working closely with the U.S. Marine Corps on exercises in both U.S. and Japanese training ranges designed to improve Japan's competency in amphibious operations.

This capability to rapidly employ the GSDF to offshore islands adds to several capabilities Japan already fields to protect its sovereignty. Its navy operates a number of surface vessels to patrol its territorial waters, in addition to maritime patrol aircraft and helicopters to provide maritime situational awareness of both surface and subsurface threats. In recent years, all these capabilities have been tested by Chinese air and maritime incursions. Japan responds to each of these by sending an appropriate air or surface vessel to escort Chinese vessels out of Japanese waters.

Japan continues to field and make new investments in ballistic missile defense capabilities. North Korea, in particular, is seen as a key missile threat for Japan, while China also has a substantial arsenal that can reach Japan. Japan and the United States have cooperated for many years on missile defense research, and Japan fields a mixture of U.S. and indigenous systems. Japan plans to expand its missile defense–capable ships to eight. Japan fields Patriot air defenses and plans to invest in the newest interceptor missiles, the PAC-3 MSE (Patriot Advanced Capability Missile Segment Enhancement).

What the U.S. Army Can Do to Help the Japanese GSDF Help Itself

The implementation of major changes in the GSDF creates a substantial opportunity for the U.S. Army to partner with the GSDF in new and productive ways. The U.S. Army has air and missile defense expertise, as well as sustainment expertise, that could benefit the GSDF as it contemplates more challenging scenarios for defending Japanese territory. In such scenarios, fielded forces may face attacks from ballistic

---

[12] Ministry of Defense of Japan, *Medium Term Defense Program (FY2014–FY2018)*, Tokyo, December 17, 2013b.

and cruise missiles, and supply lines might also come under attack. The U.S. Army is well situated to engage with Japan on operational problems such as these.

Japan's GSDF already fields a modest number of truck-mobile units equipped with indigenous anti-ship cruise missiles. This provides additional firepower to anti-ship capabilities from Japan's surface fleet, submarines, and the Japan Air SDF. At the moment, the threats Japan anticipates from China and North Korea involve low levels of force, and in neither case does Japan's Medium Term Defense Program reference the possibility of a large-scale maritime attack.[13]

The existing anti-ship forces could compose the basis for a U.S.-Japan operational concept to integrate anti-ship, surface-to-surface, and surface-to-air missile units into a combined multi-domain A2/AD capability. Such an operational concept could allow the United States and Japan to test their capabilities in exercises and training events and to expand these capabilities as required. For its part, Japan has the resources and ability to expand its land-based anti-ship capability if it believes that the threat warrants additional attention in the future.

---

[13] Ministry of Defense of Japan, 2013b.

CHAPTER FOUR
China-Philippines Relationship from the Philippines' Standpoint

In this chapter, we examine the China-Philippines relationship from the Philippines' standpoint. We start with some context on the U.S. relationship with the Philippines, including reflections on the changing relationship under Philippines President Rodrigo Duterte. We then examine some potential China-Philippines conflict scenarios, the Philippines' security posture and operational concept, and how the U.S. Army can help the Philippines help itself.

Context: U.S. Relationship with the Philippines

Although the United States closed its military bases in the Philippines more than 20 years ago, U.S. concern over regional stability and growing tensions between the Philippines and China have reinvigorated the United States' relationship with the Philippines. The two countries continue to uphold a defense treaty signed in 1951 (Mutual Defense Treaty Between the Republic of the Philippines and the United States of America). The treaty obligates each party to recognize that an "armed attack" in the "Pacific Area on either of the Parties" would be "dangerous to its own peace and safety." The treaty obligates each party to "act to meet the common dangers in accordance with its constitutional processes."[1]

---

[1] U.S. Department of State, undated.

As discussed in Chapter Two, Manila's disputes with Beijing have centered primarily on contested claims over features in the South China Sea and the fishing and mineral resources contained therein. After Vietnam, the Philippines occupy and control the most number of islands and reefs in the contested Spratly Islands. A garrison of roughly 40 Filipino soldiers guards the Spratly's second-largest island, Thitu Island, which measures roughly 47 square kilometers. (Taiwan controls the largest island, Itu Aba.) Thitu Island also features one of the Spratly Islands' few runways. Manila also controls another six islands and three reefs and exercises control of over another dozen shoals, reefs, and other generally submerged features. Thitu Island and many of the other disputed features are located roughly 200–300 nautical miles from Palawan.

Philippine military forces pose little threat to China. Philippine defense planning remains focused principally on internal security. Although Manila recognizes the growing importance of external threats and has acquired a coast guard cutter, the country still lacks adequate air defense, maritime patrol, and reconnaissance capabilities.[2]

Because of the weakness of its military forces, Manila has relied on asymmetric means to draw attention to Chinese assertive behavior and counter Chinese tactics. It has ensured access for media personnel aboard boats that resupply beleaguered outposts. In early 2013, the Philippines launched an arbitration case under the United Nations Convention on the Law of the Sea against China's South China Sea claims, in which the Philippines won a resounding victory, as described in Chapter Two.[3]

While the primary dispute between China and the Philippines centers on territory and sovereignty in the South China Sea, Manila's alliance with Washington aggravates broader issues of dispute over status and leadership. As competition between China and the United

---

[2] Evan S. Medeiros, Keith Crane, Eric Heginbotham, Norman D. Levin, Julia F. Lowell, Angel Rabasa, and Somi Seong, *Pacific Currents: The Responses of U.S. Allies and Security Partners in East Asia to China's Rise*, Santa Monica, Calif.: RAND Corporation, MG-736-AF, 2008, p. 117.

[3] Council on Foreign Relations, *China's Maritime Disputes*, Washington, D.C., 2016.

States intensifies, the Philippines, like other U.S. allies, may be viewed as proxies of U.S. power. How disputes between China and the Philippines are resolved unavoidably implicates U.S. prestige and influence. The credibility of the United States will be damaged if a crisis or clash is resolved in a manner that appears contrary to the wishes of Washington. If China and the United States were to reach a regional accommodation that appeared to go against the Philippines' preferences, Manila may begin to doubt the value of the alliance.

This relationship between the Philippines and the United States, and the Philippines and China, may now be fundamentally changing. While visiting China, President Duterte stated that he was moving toward China and that the United States had "lost," stating: "I announce my separation from the United States. I have separated from them. So I will be dependent on you for all time. But do not worry. We will also help as you help us."[4]

These and other statements by Duterte signal a profound change in public rhetoric and perhaps in the future course of Chinese and Philippine actions. For example, there are early reports that the Chinese forces occupying the Scarborough Shoal are allowing Filipinos to fish there again without interference.[5] Further, Duterte has stated that he wants the Philippines "free of the presence of foreign military troops" and intends to end maneuvers and wargames alongside the U.S. military.[6]

Potential China-Philippines Conflict Scenarios

Only time will tell whether this new relationship between the Philippines and China will continue and whether close diplomatic and

---

[4] Associated Press, "Philippine President Duterte Announces Separation from U.S.," October 20, 2016.

[5] Emily Rauhala, "Philippines Says China Has Stopped Chasing Fishermen from Contested Shoal," *Washington Post*, October 28, 2016.

[6] Simon Denyer, "Philippine Leader Duterte Now Wants U.S. Troops Out 'in the Next Two Years,'" *Washington Post*, October 26, 2016.

military cooperation between the Philippines and the United States will significantly diminish or end. For the purpose of our analysis, we describe future courses of events that might proceed from either this new China-Philippines relationship or the more-traditional China-Philippines-U.S. relationships.

As its power grows, China might feel inclined to exert its influence in a manner that directly involves the Philippines in two ways. First, China could seek to gain even greater control of important maritime and land features in the South China Sea claimed by the Philippines. Second, China could seek opportunities to humble those countries that oppose the extension of Chinese power. The most important of these would be allies of the United States. Thus, to the degree that the Philippines chooses to continue its relationship with the United States, China could target the Philippines as a proxy of U.S. power to score points against its strategic rival.

These two imperatives would likely underpin any crisis or military confrontation between China and the Philippines. While it is unlikely that Beijing would choose military aggression as its first option to achieve its goals, it could end up in conflict through miscalculation. In addition, a China determined to break up U.S. authority and leadership in Asia could target the Philippines as a proxy for U.S. power, further raising the risk of conflict. We explore both types of scenarios below.

Scenarios of Miscalculation

The first type of scenario consists of an unintended escalation arising from a crisis between China and the Philippines over a particular disputed reef or island. In this case, China remains focused on a "salami slicing" strategy of pressure through harassment by coast guard, fishing, and other vessels; the continued reclamation of land for its reefs; enhanced military presence; vigorous prosecution of China's case through media and legal channels; economic retaliation through reductions in imports and other measures; and subtle threats to the Philippines and other regional powers about the potential risks of challenging Beijing.

Convinced of the superiority of its approach, Beijing might intuit that such neighbors as the Philippines or Vietnam might attempt to derail China's incremental consolidation of control through dramatic action. To forestall such a possibility, China would backstop with naval and air power its largely civilian efforts to wrest control of the South China Sea features. Even today, China carries out whole-of-government efforts to consolidate control of disputed regions, with a substantial military presence providing security.

The most likely trigger for crisis in this scenario would thus begin with an action by the Philippines to halt some nonmilitary action by the Chinese to challenge Philippine control of some contested feature. The most likely target would be another unoccupied reef or shoal, over which the Philippines has traditionally exercised control. In something of a repeat of the Scarborough Shoal situation, Chinese authorities could seek to exploit a misstep or overreaction by the Philippines to justify a further consolidation of control over the disputed reef. If the Philippines refused to back down, or even stepped up efforts to defend its reef, the situation could quickly escalate, possibly even resulting in an exchange of fire and lives lost. Intense international pressure to restore peace would likely propel China, the Philippines, and the United States to find ways to deescalate the situation.

Washington's words and actions would have a large influence in determining the trajectory of subsequent clashes between Beijing and Manila. A disengaged or acquiescent United States would give Manila little hope of prevailing against Beijing in any confrontation and probably undermine its resolve, depriving it of real leverage. If the United States appeared unwilling to risk a confrontation over the issue, the extreme imbalance in power would leave Manila with little recourse but to accept the change in the status quo, much as it did in the wake of China's acquisition of Mischief Reef and Scarborough Shoal. Moreover, an emboldened China could be tempted to repeat its success. Seeing little opposition, China could even dispense with the time-consuming, laborious process of operating through various civilian fishing and coast guard proxies and seize control outright of more reefs and islands with military force.

A humiliated Philippines that witnessed a string of defeats and loss of control over long-claimed islands and features could, in theory, feel desperate enough to use available military capabilities to strike Chinese platforms to stem the erosion of its position. Such an act would almost certainly cause China to strike back—raising questions about how the United States should respond. Assuming that U.S.-China relations remained steady, and that the United States and the Philippines retain their alliance, Beijing would push Washington not to support its ally. This would also serve China's objectives of weakening U.S. influence and alliances in the Asia-Pacific region.

Scenarios of Chinese Aggression
Because of its formal alliance with the United States, the Philippines makes for a risky target of unprovoked aggression. Military attacks on Philippine-occupied territory, such as Thitu Island, would be most plausible under conditions in which China felt that its peaceful strategy to control the region had failed and that time was no longer on China's side, or if China felt that opposition had grown feeble and the opportunity proved ripe.

One scenario for aggression is one in which China grows pessimistic about its future because of a severe weakening of its strategic position. If Beijing concluded that inaction risked even greater loss and disaster than some sort of military action, then it could opt for a risky course of action, knowing full well that the United States could become involved. In this case, China would likely design a military course of action to maximize its prospects for success against the Philippines and minimize the odds of success by an intervening U.S. military force. The high-risk move would gamble on the idea that the United States would back away from conflict with China, allowing Beijing to make territorial gains, intimidate the Philippines, and improve its strategic position.

The largest drawback to this operation would be miscalculation. Even if China successfully evicted Philippine forces from a given South China Sea feature, such as Thitu Island, China would be challenged to sustain forces on the occupied feature in the face of a concerted U.S.-Philippines military intervention to retake the feature. Additionally, once U.S. and Philippine forces engaged Chinese forces, the risk of escalation would increase dramatically. Chinese naval forces would likely attempt to interdict the U.S. forces, but these platforms, lacking air cover, would be highly vulnerable. It would be safer for China to make a point by attacking Philippine forces and then withdrawing. But even this course of action poses significant risk of escalation, because it would almost certainly result in the invocation of the mutual defense treaty.

A more promising option for China would be to wait for the opportunity when it seemed that the Philippines lacked any hope of relief from its U.S. ally and when it seemed that no one else appeared willing or capable to help the Philippines defy China's seizure of islands and reefs.[7] Through superior military capabilities, the PLA could seize Thitu Island and other features. To make this successful, however, China would need to be certain that the United States would do nothing to support its ally, a doubtful proposition as long as the alliance remains in force. Although Washington may in most circumstances be reluctant to risk war over Manila's South China Sea claims, it would likely feel intense pressure to provide some level of assistance to help its ally protect its claims and ward off further Chinese predations in a situation of outright aggression.

Addressing the China Threat from the Philippines' Standpoint

The Philippines' Security and Military Posture in Response to the China Threat

The Philippine Army, the largest service in the country by far, has trained and organized for counterinsurgency (COIN), specializing in

---

[7] See Chris Mirasola, "Water Wars: The Philippines Pivots Towards Beijing," *Lawfare*, September 30, 2016. Duterte recently commented that he would open an alliance with China and Russia and expressed doubt that the United States would aid the Philippines if attacked. Although it is premature to draw conclusions from this outburst, it may be that the Philippines already doubts U.S. security commitments. It remains to be seen how President Trump will view the U.S. alliance with the Philippines and if and how he might wish to push back against Chinese ambitions in the western Pacific.

small-unit (battalion and below) operations.[8] The picture that emerges from the information available about Philippine military modernization is that the country is clearly reorienting toward conventional combat operations. For many years, the Philippine armed forces neglected air defenses, anti-ship weapons, and anti-submarine capabilities. All those gaps are now being addressed, at least to some extent; however, the process of retooling the Philippine military toward conventional combat is just beginning and will take years. Despite its close relationship with the United States military, the Philippine military is looking at multiple sources for new equipment, exemplified by the emerging deal with Israel for air defenses and the purchase of both fighters and frigates from South Korea.

The Philippine Army has several hundred armored personnel carriers (APCs) (both tracked and wheeled) but has no main battle tanks and fewer than ten light tanks. Neither the nation's air force nor the army has air defense weapons. The Philippine Navy has focused on coastal operations to support army and marine COIN missions. Some of the nation's navy patrol craft are capable of offshore operations in the South China Sea, but their armament is oriented toward anti-piracy and law enforcement-type missions in a low-threat environment. Importantly, the Air Force has no modern fighters, and the Navy and Air Force do not have ASMs.

In 2015, the outgoing Philippine president, Benigno Aquino, pledged to increase defense spending by 25 percent for 2016 to support a $1.8 billion, five-year modernization plan undertaken in 2013. President Duterte has largely upheld these spending commitments.[9] According to a Philippine news report in 2016, the Philippine military is expected to add its first landing platform dock from Indonesia, two C-130 aircraft from the United States, an additional two TA-50 aircraft from South Korea (the first two arrived in December 2015),

---

[8] Interviews with U.S. Army Special Operations personnel, Fort Bragg, N.C., 2014.

[9] Christine C. Cudis, "AFP Budget to Be Spent on Modernization of External Defenses," *PhilStar Global*, May 27, 2016.

and two light-transport aircraft from Indonesia.[10] Budgeted purchases through 2017 include[11]

- 12 FA-50 light fighters (from South Korea)
- six close air support aircraft (vendor still to be determined)
- two Navy frigates (from South Korea)
- ten Patrol boats for the Navy (vendors still to be determined)
- 21 transport helicopters (from the United Kingdom and Canada)
- two anti-submarine helicopters (vendors to be determined)
- two sealift, amphibious ships (from an Indonesian vendor)
- roughly 150 M-113 APCs (from U.S. Army stocks)
- air defense radars (from Israel)
- 12 launchers of ground-launched ASMs[12] (these systems, along with appropriate radars for target acquisition and tracking, will be under the control of the Philippine Army;[13] a number of possible suppliers are currently being examined).

The Philippine military has been in discussion with Israel about the possible purchase of air defense systems from Rafael Advanced Defense Systems.[14] Given the almost total lack of an air defense capability in the Philippine military, this would be a step toward filling an important gap.[15] In addition to the air defense purchase from Israel,

---

[10] Frances Mangosing, "More Ships, Planes, Combat Gear for AFP in 2016," *Inquirer.net*, January 1, 2016.

[11] Agence France-Presse, "President Says Philippines to Spend $1.8 Billion on Military Modernization," *Defense News*, December 21, 2015; *Defense News*, "Philippines Hikes Defense Budget 25%," July 21, 2015.

[12] Congress of the Philippines, An Act Amending Republic Act No. 7898. Establishing the Revised AFP Modernization Program and for Other Purposes (Armed Forces of the Philippines New Modernization Act), Republic Act No. 10349, Manila, 15th Congress, 3rd regular session, December 11, 2012.

[13] *Zambo Times*, "PH to Acquire Shore-Based Missile System," December 4, 2013.

[14] Carmela Fonbuena, "PH Finalizing Air Defense Radar Deal with Israel," *Rappler*, July 9, 2014.

[15] *Manila Standard Today*, "PH Plans to Tap Israel for Missile Launchers," June 15, 2013.

the Philippine government is also considering buying the Raytheon HAWK-5 SAM.

The Philippine Navy is also considering adding an ASM capability when it purchases new versions of its Multi-Purpose Attack Craft that would be able to operate in the South China Sea but only in good sea conditions because of their small size. These small but fast (up to 45 knots) vessels could mount short-range ASMs in addition to the automatic weapons that they currently mount.

The Philippines' Current Operational Concept in Response to the Chinese Threat

Dealing with China is a new challenge for the Philippine military—and clearly still a work in progress. Prior to the early 2000s, the ability of the PLA to threaten the Philippines was limited, and with its longstanding defense ties with the United States, the Philippines did not feel threatened by China. Indeed, the Philippines felt secure enough in the early 1990s to refuse to renew the leases on the two major U.S. bases in the islands—Clark Air Force Base and the Navy's Subic Bay facilities—both on the main island of Luzon.

The Philippine attitude toward China, however, is now changing rapidly. The various Philippine military modernization programs noted above are mostly oriented toward conventional combat operations, with the Chinese now clearly seen as the main potential opponent. The Philippine military leadership knows that it is at a considerable disadvantage compared with China. In September 2014, General Gregorio Catapang, the senior officer of the Philippine armed forces, stated that the Philippine military needed an additional $10 billion in modernization funds to have the kinds of capabilities it needs to counter Chinese moves in the region.[16]

From a policy perspective, the increased emphasis on China is demonstrated in the April 2014 Enhanced Defense Cooperation Agreement (EDCA) signed between the United States and the Philippines. The EDCA is intended to improve overall defense ties between the two

---

[16] Karl Lester and Clarissa Batino, "Outgunned Philippine General Seeks Arms Upgrade as China Expands," *Bloomberg News*, September 5, 2014.

allies. This agreement establishes a formal framework for an increased U.S. military presence in the islands, although this new arrangement will not result in permanent U.S. bases in the Philippines along the lines of Clark Air Force Base and Subic Bay. Key provisions of the EDCA include the following:

- The preamble states: "Both parties share an understanding for the United States not to establish a permanent military presence or base on the territory of the Philippines."
- The preamble concludes: "All United States access to and use of facilities and areas will be at the invitation of the Philippines and with full respect for the Philippine Constitution and Philippine laws."
- Article III states that U.S. forces and their contractors are authorized to undertake "[t]raining, transit, support and related activities; refueling of aircraft; bunkering of vessels; temporary maintenance of vehicles, vessels and aircraft; temporary accommodation of personnel; communications; prepositioning of equipment, supplies, and materiel; deploying forces and materiel; and such other activities as the Parties may agree."
- Article IV states that the United States is granted the right to store prepositioned materiel, "to include, but not limited to, humanitarian assistance and disaster relief equipment, supplies, and materiel."[17]

The EDCA, therefore, envisions a rotational, as opposed to permanent, U.S. military presence in the islands. Importantly, the United States will be able to preposition a variety of equipment, including munitions, in the Philippines, which should help facilitate the rapid expansion of U.S. capabilities in the western Pacific in the event of either a humanitarian or a military crisis.

Despite the ongoing reorientation of parts of the Philippine military toward conventional operations, much of the Philippine Army will

---

[17] Carl Thayer, "Analyzing the US-Philippine Enhanced Defense Cooperation Agreement," *The Diplomat*, May 2, 2014.

likely remain focused on COIN and internal security. There is no way the Philippines can, on its own, come close to matching the military potential of China. Filipinos know this, which is why they are increasing their defense ties with the United States, clearly their most powerful potential partner in the region and a nation that has deep historical ties to the Philippines. Contributing to this willingness to partner with the Americans is the fact that Filipinos have a very positive view of the United States. A 2014 Pew Research poll of 43 nations showed a 92-percent favorable rating of the United States in the Philippines—the highest rating of all the countries in the survey.[18]

However, the trend toward closer U.S.-Philippine relationships can change suddenly and with very little warning. In rapid succession, Duterte threatened to end cooperation with U.S. forces against the Abu Sayyaf, end U.S.-Philippine maritime exercises, and shift diplomatic emphasis to closer relations with China.[19] Perhaps the most serious of Duterte's threats is to end the EDCA with the United States.[20] Although it remains too early to conclude how seriously to take Duterte's warnings, his statements do underline the risks of assuming that these relationships will not change.

What the U.S. Military Could Do to Help the Philippines Help Itself

The U.S. military has had ties to the Philippines for more than a century. Even after Clark Air Force Base and Subic Bay were returned to Philippine control in 1991, the two countries maintained an alliance relationship that has included regular military exercises, such as the PACOM-sponsored annual Balikatan exercise that includes U.S. Army

---

[18] Pew Research Center, *Global Opposition to U.S. Surveillance and Drones, but Limited Harm to America's Image: Many in Asia Worry About Conflict with China*, Washington, D.C., July 14, 2014.

[19] Mirasola, 2016.

[20] Kevin Lui, "Philippine President Rodrigo Duterte Threatens to End Defense Pact with the U.S.," *Time*, October 3, 2016.

participation. Balikatan and other combined exercises help improve interoperability between the two nations' armed forces.[21]

Since the closing of Clark Air Force Base and Subic Bay in 1991, the U.S. Special Operations personnel associated with Operation Enduring Freedom–Philippines (OEF-P) (fighting radical extremist groups in the southern islands) have composed the largest, most regular U.S. military presence in the Philippines. Starting in December 2001, PACOM established a rotational, but essentially constant, special operations forces (SOF) presence in the southern Philippines.[22] Most of the U.S. SOF personnel are either U.S. Army Special Forces or U.S. Navy SEALs. They operate closely with all elements of the Philippine armed forces, providing training, mentoring, and advice and also serving as the conduit of U.S. intelligence information directly into the Philippine armed forces. Although the OEF-P is a SOF-focused activity at the lower end of the spectrum of conflict and the number of U.S. SOF personnel in the Philippines is usually well under 500 at any time, the experience that OEF-P provides has contributed to the interoperability of the two countries' armed forces.[23]

Thus, the new opening in the Philippines provided by the EDCA, the regularly occurring military exercises (such as Balikatan), and the ongoing OEF-P relationship all form a foundation for increased interaction with the Philippine military, including by the U.S. Army. The Philippine armed forces are developing concepts of operations to deal with various types of sovereignty challenges. Coordinating U.S. and Philippine military approaches to deal with minor provocations in the South China Sea or a potential military confrontation should be the basis for combined operational concepts. As one example, the United States would be well positioned to contribute theater-wide intelligence, surveillance, and reconnaissance (ISR) to these concepts.

The EDCA provides for prepositioning U.S. military equipment and supplies in the Philippines—a provision that could significantly

---

[21] *Global Security*, "Exercise Balikatan: Shouldering the Load Together," October 4, 2016.

[22] *Global Security*, "Operation Enduring Freedom-Philippines," May 7, 2011.

[23] Discussions with former U.S. Army SOF personnel who participated in OEF-P, Washington D.C., November–December 2014.

improve the U.S. military's ability to quickly grow its military capabilities in the western Pacific in the event of a crisis. Although U.S. military units will not be permanently assigned to the Philippines, prepositioned stocks could be located in a number of key points in the islands. The U.S. Air Force and the U.S. Navy are already considering positioning equipment in the Subic Bay region and in several Philippine airports and military air bases. The U.S. Army, working closely with PACOM headquarters and the Philippine military, should identify suitable locations and the types and quantities of equipment and supplies that should be considered for prepositioning.

Given the tendency of this region to suffer major natural disasters, a considerable portion of prepositioned stocks could be oriented toward humanitarian and disaster relief operations. Certain types of munitions—in particular, air defense weapons, such as Patriot missiles—are difficult to quickly transport by air on short notice. It may be possible to locate a judicious number of this class of munitions in the Philippines. The types of equipment the U.S. Army decides to recommend for prepositioning in the Philippines should be based on joint considerations among U.S. forces, such as possible Army roles in the Air Force's still-emerging adaptive basing concepts, and on the ability of Philippine military to provide key capabilities, thus eliminating or minimizing need to deploy U.S. assets to perform that function.

CHAPTER FIVE
China-Taiwan Relationship from Taiwan's Standpoint

In this chapter, we examine the China-Taiwan relationship from Taiwan's standpoint. We start with some context on the U.S. relationship with Taiwan and then examine some potential China-Taiwan conflict scenarios, Taiwan's security posture and operational concept, and how the U.S. Army can help Taiwan help itself.

Context: U.S. Relationship with Taiwan

Taiwan was a U.S. ally during the Cold War, but as the United States began to move closer to China to balance against the Soviet Union in the early 1970s, it became clear that the normalization of the U.S.-China relationship would change the nature of the United States' relationship with Taiwan. When the United States ultimately moved to shift official diplomatic recognition to China, it also began the process of withdrawing its forces from Taiwan and abandoning the Sino-American Mutual Defense Treaty.

The United States frequently reaffirms its commitment to the One China policy based on the Three Joint Communiqués (the 1972 Shanghai Communiqué, the 1979 Communiqué on the Establishment of Diplomatic Relations, and the 1982 U.S.-PRC Joint Communiqué) and the 1979 Taiwan Relations Act (TRA).[1] Indeed, the Three Joint

---

[1] See, for example, White House, "Remarks by President Obama and President Xi Jinping in Joint Press Conference," Office of the Press Secretary, November 12, 2014b.

Communiqués provided the basis for the normalization of the U.S.-China relationship and continue to play an important role in U.S.-China relations. The TRA, meanwhile, provides the framework for robust, but unofficial, diplomatic, economic, and security relations between the United States and Taiwan. The TRA is also intended "to help maintain peace, security, and stability in the Western Pacific." It underscores an enduring U.S. interest in ensuring that Taiwan's future is determined by peaceful means and states that it is U.S. policy "to consider any effort to determine the future of Taiwan by other than peaceful means, including by boycotts or embargoes, a threat to the peace and security of the Western Pacific area and of grave concern to the United States." Moreover, the TRA pledges that the policy of the United States is "to provide Taiwan with arms of a defensive character" and "to maintain the capacity of the United States to resist any resort to force or other forms of coercion that would jeopardize the security, or the social or economic system, of the people on Taiwan."[2] The TRA does not obligate the United States to assist Taiwan militarily in any contingency. Any Chinese military action against Taiwan, however, would quickly become an extremely serious challenge to U.S. security commitments and threaten the stability and security of the region. This would add pressure on Washington to take some sort of action.

The United States continues to have an interest in ensuring that any future resolution of cross-Strait issues is peaceful and accords with the will of Taiwan's 23 million people. In short, even though the United States no longer maintains a formal diplomatic relationship or defense treaty with Taiwan, the TRA and U.S. policy make it clear that Washington would see any use of force against Taiwan as a grave threat to regional peace and stability.

Taiwan rejects the framework of a "one China, two systems" proposal by China, because this would imply Taipei's recognition of Beijing's authority over it. However, Taiwan's disagreement with China on its political status is complicated by the various positions held by the political parties on the island. Taiwan's leaders agreed in 1992

---

[2] Public Law 96-8, Taiwan Relations Act, 96th Congress, Washington, D.C., January 1, 1979.

with China that both sides belong to "one China" but that China and Taiwan can interpret that principle however they wish, an agreement regarded as the 1992 Consensus. Although the preceding president, Ma Ying-jeou of the Kuomintang Party, supported the 1992 Consensus, the current president, the Democratic Progressive Party's Tsai Ing-wen, has refused to embrace it. Tsai, however, has so far avoided the behavior that highly antagonized Beijing and that was characteristic of President Chen Shui-bian, also of the Democratic Progressive Party, from 2000 to 2008.[3]

Taiwan's strategy for maintaining its de facto sovereignty without provoking a war with China depends on its ability to maintain a strong economic and international presence and robust security ties with the United States. Taipei seeks to maintain diplomatic relations with as many countries as possible to raise the diplomatic cost to China of aggression, although this number has dwindled to a handful of islands and small countries that maintain the position primarily for economic benefit. Taiwan has also tried to increase its profile in international events and organizations, such as the World Health Organization, but China's considerable diplomatic clout allows it to greatly influence the terms of Taiwan's official presence in such bodies. Economic connections remain another key mechanism of restraint. As one of the world's most prosperous economies, Taipei maintains strong trade links with countries around the world.

Taiwan continues to rely heavily on an implied security commitment from the United States to deter China. In addition to arms sales, Taiwanese and U.S. officials carry out high-level meetings on security policy and professional military exchanges. U.S. military personnel have also observed Taiwan's military exercises.[4]

In 2012, Andrew Yang, one of Taiwan's leading national security analysts and a former minister and vice minister of national defense,

---

[3] Tom Philips, "Taiwan's New President Tsai Ing-wen Vows to Reduce Dependence on Beijing," *Guardian*, May 20, 2016.

[4] Alexander Huang, "The United States and Taiwan's Defense Transformation," *Taiwan-U.S. Quarterly Analysis*, Brookings Institute, February 2010.

stated: "Even when we have peaceful cross-Strait relations, [as we do] right now, Beijing is still holding a big stick. So we have to be on alert."[5]

Potential China-Taiwan Conflict Scenarios

Although Taiwan and China have made remarkable progress in forging a more stable and peaceful relationship in recent years, the fundamental political issues that divide them remain unresolved, and the future of the cross-Strait relationship remains uncertain.[6] China could choose to use military force to compel unification or deter further provocations if Taiwan's leadership declared independence or otherwise pursued actions that appeared to elevate the possibility of de jure independence. The risk of conflict would increase significantly if these developments on Taiwan took place within the context of an intensifying rivalry between China and the United States, thus removing the United States as a potential stabilizing influence on any such situation.

Here, we examine two potential conflict scenarios: (1) crisis-escalation scenarios and (2) compel-unification scenarios.

Crisis-Escalation Scenarios

The most likely scenarios for the near future remain crises in which China employs force for primarily punitive or deterrent purposes. In these cases, China's aversion to major escalation could offer the United States opportunities to reach out to China and Taiwan to deescalate the situation.

In a crisis that Beijing regarded as serious but not grave enough to require the lethal use of force, it could resort to a number of demonstrations and other military or paramilitary activities to signal its displeasure and warn Taipei against taking any action that might be seen as provocative. Chinese actions could be augmented by intense psycho-

---

[5] *Defense News*, "Interview: Nien-Dzu Yang, Taiwan's Vice Minister of Defense—Policy," November 12, 2012.

[6] Richard Bush, *Uncharted Strait: The Future of China-Taiwan Relations*, Washington, D.C.: Brookings Institution Press, 2013.

logical and information warfare activities to coerce Taipei into making decisions favorable to Beijing's interests. China's missile tests and military exercises during the cross-Strait tensions in 1995 and 1996 may be viewed as this sort of deterrence operation.

A more serious move would be to employ military force in a limited way to intimidate, deter, or punish Taipei. Options could include a computer network attack, a missile strike on Taiwanese military assets, or possibly even a limited missile strike on the island. With this sort of operation, there could be little warning time. China, however, would also then be poorly positioned to carry out large-scale follow-on attacks to exploit the effects of surprise. A repetition of such incidents with no clear resolution would increase the likelihood that China would begin preparations for major operations as an escalation option.

Compel-Unification Scenarios
Less likely, but more dangerous, are scenarios in which China initiated major combat operations to compel Taiwan's unification. The trigger could be any of the conditions listed in China's National Anti-Secession Law. The risk of China pursuing this type of military operation would increase following a series of cross-Strait crises, an intensification of U.S.-China rivalry, or an increasingly pessimistic Chinese assessment about the possibilities of achieving peaceful unification.

PLA literature describes a number of types of campaigns that could be relevant to such an escalating crisis scenario. China's "conventional missile attack campaign" would involve a "series of conventional missile attacks" aimed at the enemy's "important targets."[7] The PLA Second Artillery Force would take the lead role in this campaign, but PLAAF and PLAN units could also play important roles. Such a campaign could be executed as a stand-alone campaign for coercive purposes or to help China seize air, sea, and information superiority in support of other campaigns, such as the "joint blockade campaign" or "joint island landing campaign."[8] Missile attacks could inflict great

---

[7] Yu Jixun, ed., *The Science of Second Artillery Campaigns*, Beijing: PLA Press, 2004.

[8] See Michael Chase, Jeffrey Engstrom, Tai Ming Cheung, Kristen Gunness, Scott Warren Harold, Susan Puska, and Samuel K. Berkowitz, *China's Incomplete Military Transformation:*

havoc with minimal warning, and there is little that can be done to protect against such strikes. Nevertheless, this course of action would be unattractive for at least two reasons. First, missile attacks alone are unlikely to compel Taiwan's capitulation—and mounting military and civilian casualties from missile bombardment could harden Taiwan's resolve against Beijing. Second, missile attacks could inflict considerable damage on the country, which China would have to reconstruct should it succeed in annexing the island.

The PLA's joint blockade campaign is a "protracted campaign" that would be undertaken to "sever enemy economic and military connections" with the outside world and thereby compel the enemy to submit to China's demands.[9] PLA literature suggests that this campaign is envisioned as including conventional air and missile strikes and information and electronic attacks against the enemy to shatter its ability to resist the blockade. The drawback to this military campaign is that it lacks a clear mechanism to compel Taiwan's capitulation. The evidence remains weak that an air and naval blockade alone would compel a country to capitulate. The effect is often the opposite—a hardening of resolve on the part of the blockaded party. Moreover, this course of action leaves the PLA forces vulnerable for an extended period, enabling Taiwan to gain sympathy among friendly nations and for the United States to marshal forces able to escort merchant shipping and perhaps also attack blockading naval and air platforms.

The PLA joint island landing campaign would be designed to "seize and occupy a whole island or important target." To successfully accomplish this objective, the PLA must first successfully cross the

---

Assessing the Weaknesses of the People's Liberation Army (PLA), Santa Monica, Calif.: RAND Corporation, RR-893-USCC, 2015. A coercive or demonstrative use of conventional missile firepower, such as the series of launches China conducted during the 1995–1996 Third Taiwan Strait Crisis, could also be related to this type of campaign and could escalate to a conventional missile attack campaign if intimidation short of that level fails to achieve the desired objectives.

[9] Zhang Yuliang, ed., *The Science of Campaigns,* Beijing: National Defense University Press, 2006, p. 292.

Taiwan Strait, destroy Taiwan's defenses, and secure a beachhead.[10] The joint island landing campaign provides the most convincing military path to unification. In occupying and controlling the seat of government, China could ensure that the island's leaders accept whatever terms Beijing requires. Worse yet, Taiwan's ability to contest an invasion is atrophying even as China improves its ability to carry out so complex an operation.[11] Even so, an opposed amphibious assault is a high-risk operation, especially given the limited lift available to China. Moreover, a large-scale amphibious assault by nature would likely lead to war—one that has a high risk of drawing in large-scale U.S. involvement.

Addressing the China Threat from Taiwan's Standpoint

Taiwan's Security and Military Posture in Response to the China Threat

Taiwan's strategy for addressing the threat from China relies on conventional capabilities and readiness to deter an invasion and other attacks. Taiwan recognizes that its military would be operating at a disadvantage against China. If faced with an invasion, Taiwan expects a substantial initial onslaught. The first of five identified Taiwanese military strategic missions is "Resolute Defense to Ensure the Security of National Territories," which involves "surviv[ing] the enemy's first strike, avert[ing] decapitation, and maneuver[ing] forces to counter strike and sustain operations."[12] In short, Taiwan seeks a military that can withstand a massive first punch and remain a functional fighting force. Taiwan's capability development emphasizes increasing the precision firepower and mobility of its force to "enable the [Republic

---

[10] Bi Xinglin, ed., *Campaign Theory Study Guide,* Beijing: National Defense University Press, 2002, pp. 225–226.

[11] See, for example, Michael J. Lostumbo, David R. Frelinger, James Williams, and Barry Wilson, *Air Defense Options for Taiwan: An Assessment of Relative Costs and Operational Benefits,* Santa Monica, Calif.: RAND Corporation, RR-1051-OSD, 2016.

[12] Ministry of National Defense of the Republic of China, *National Defense Report: 2013,* Taipei, October 2013b, p. 35.

of China] Armed Forces to take advantage of tactical situations and reverse unfavorable conditions."[13]

Taiwan has a multifaceted concept for ground defense requiring layered defensive capabilities to enable its ground force to strike an invasion force before it reaches Taiwan, strike it on the landing beaches, and counterstrike as the invasion force moves inland. The ground defense concept also describes the need to counter enemy airborne forces. It prudently emphasizes the importance of force preservation in a complex battlefield involving precision fires. To counter the attackers at sea, the Taiwanese Army would use multiple strike systems: a multiple launch rocket system (MLRS), unmanned aerial systems (UAS), and attack helicopters. To prevent the enemy from establishing a lodgment on the beach, Taiwan needs anti-armor missiles, short-range anti-armor rockets, and modern tanks. Taiwan plans for the active-duty forces to conduct strike and maneuver operations, while the reservists conduct "homeland defense" operations, which appears to refer to both static defenses against maneuver forces and rear-area security.[14] Taiwan identified a need to enhance mobility and fires coordination. It still sees a role for light and medium tactical vehicles and seeks self-propelled artillery. Taiwan has several garrisoned offshore islands, so the Taiwanese Army also seeks an amphibious capability to support those garrisons.

Taiwan's Current Operational Concept in Response to the Chinese Threat

Taiwan clearly articulates a defense concept and outlines the needed capabilities to support the concept. The concept aspires to joint operations and specifically calls for layered defense capabilities. Some of the aspirations appear rather unobtainable given the current threat and Taiwan's current military spending. In particular, Taiwan has not adjusted to its loss of a clear qualitative advantage over PLA capabili-

---

[13] Ministry of National Defense of the Republic of China, *2013 Quadrennial Defense Review*, Taipei, March 2013a, p. 41.

[14] Ministry of National Defense of the Republic of China, 2013a.

ties; thus, in most domains, China fields more-numerous and more-competent capabilities.

Beyond the recalibration of Taiwan's military goals, Taiwan needs greater prioritization in its defense program. Taiwan's published Quadrennial Defense Review and its National Defense Strategy are not resource-constrained documents. Taiwan's strategy should, but does not, match realistic budget projections. A corollary to the problem of wishing for too many new systems is keeping too many old ones and not divesting systems that are no longer able to deliver capability against the PLA.

Taiwan has historically relied on U.S. defense technology, but it has increasingly turned to indigenous design and production, which has allowed Taiwan to field several new missile systems: (1) a subsonic and a supersonic anti-ship cruise missile (Hsiung Feng [HF] II and III), (2) a land attack cruise missile (HF-2E), (3) silo-based and road-mobile short-range ballistic missiles (Tien Chi and Qingfeng), and (4) a ground-based air defense system (TK 1/2/3). Reportedly, Taiwan is also developing a new longer-range ground-launched cruise missile, the Yun Feng.[15] In general, Taiwanese Army modernization has lagged in recent years compared with its other services. A wheeled APC, the Clouded Leopard, has been designated a replacement for Taiwan's M-113s.[16]

When Taiwan is able to bring new systems into its inventory, it will need a well-trained force to use them to their utmost capacity. As Taiwan transitions to an all-volunteer force, the hope is that active-duty forces will stay longer and be better trained and thus more competent than was the case under a conscription system; however, Taiwan will have to rely on reservists to compensate for the much smaller force that Taiwan will then have. The amount of initial training and periodic refresher training currently planned for the reserve force appears too short to meet those competency objectives.

---

[15] Wendell Minnick, "Taiwan Working on New 'Cloud Peak' Missile," *DefenseNews*, January 18, 2013.

[16] Claire Apthorp, "Light Armoured Vehicle Procurement," *Defence Review Asia*, October 26, 2011.

The PLA now fields impressive systems in all military domains, giving it the potential to find and efficiently attack many of Taiwan's defensive systems. Part of what makes PLA systems so lethal is their potential ability to use sophisticated kill chains that involve airborne surveillance capabilities cuing a variety of precision long-range fires. Taiwan has rightly identified increasing firepower and agile maneuver as two key components to a successful defense strategy. In addition, having the capability to prevent target quality surveillance of its forces is a key enabling capability for Taiwan. The most survivable way to accomplish this in the face of PLA ballistic and cruise missile threats is from ground-based SAMs, but it is difficult to know how much Taiwan is investing in such capabilities.

What the U.S. Army Could Do to Help Taiwan Help Itself

Taiwan faces numerous challenges—a formidable adversary; limited defense spending; a resetting after a rapid downsizing and switch to an all-volunteer force; international political isolation; and limitations on its relationship with its strongest partner, the United States. Despite these challenges, the U.S. Army could help Taiwan to improve, particularly in the areas of planning, prioritizing potential modernization investments, and unit-level training.

The fast pace of PLA modernization has made it difficult for Taiwan's planners to respond with concepts and capabilities that match the quickly evolving threat. Taiwan's Quadrennial Defense Review directs its army to develop capabilities to interdict PLA forces as they cross the Taiwan Strait and when they land on the beach. Taiwan already has several precision-fire systems that could be used to interdict transiting forces, but, to be effective, these systems need to be networked with cuing systems that are survivable in the face of substantial threats from the air.

The size and the scope of the cross-Strait interdiction problem suggests that Taiwan cannot count on being able to prevent at least some PLA forces from landing on Taiwan. Although Taiwan has long contemplated this as a possibility, what has changed in recent years is the amount and diversity of PLA precision fires that will be supporting such operations. Although the U.S. Army has not faced an adversary with such capabilities or a situation quite like the invasion that Taiwan plans for, the U.S. Army could help Taiwan in developing plans and tactics to deal with the highly complex challenges associated with engaging PLA forces as they land.

Taiwan cannot afford to make big mistakes in selecting its defense systems. Taiwan's Army has enjoyed few major acquisitions in recent years, with the Apache helicopters being the primary exception. Keys to effective Taiwanese operations will be a combination of firepower, short-range air defense, passive defensive measures, and maneuver concepts to both exploit opportunities to attack and allow disengagement when the tactical conditions are very unfavorable. Consultations with the U.S. Army could help Taiwan chart an effective direction, given the budgetary and political constrains under which it operates.

Taiwan's armor and artillery forces are antiquated compared with those of the PLA, thus making it urgent for Taiwan to modernize them. Taiwan already has some advanced firepower systems, but it could benefit from further investments in such systems as the Army Tactical Missile System (ATACMS), High-Mobility Artillery Rocket System (HIMARS), and longer-range artillery with guided munitions rounds. The U.S. Army could help Taiwan assess whether its combat vehicles, such as the M60 tank and CM-32 APC, provide the right combination of mobility and protection against the threats the PLA is fielding. Taiwan should also consider increasing the density of antitank weapons in its maneuver forces.

Taiwan has recently invested in wide-area air defense systems, including the Patriot PAC-3 and the indigenous TK-III systems. Still, the volume of air threats suggests that further investments in a layered ground-based air defense system could better protect ground forces as they maneuver to engage enemy units. The U.S. Army is developing such a system, the Indirect Fire Protection Capability–Increment 2 (IFPC-2), which is discussed in Chapter Seven, and uses Sentinel radars and a variety of ground-launched air-to-air weapons.

Once the defensive concepts to engage landing PLA forces are refined, Taiwan's training and exercise programs should be revised to incorporate these new concepts. The U.S.-Taiwan military relationship continues to have substantial limitations; for instance, the U.S. military does not conduct exercises on Taiwan, but Taiwan's Air Force for years has kept F-16 fighter aircraft in the United States to allow Taiwan's pilots to train in the United States. The U.S. Army should consider whether a comparable arrangement with Taiwan's Army could be implemented, with the U.S. Army hosting unit-level exercises at U.S. training facilities. These exercises could start with small units at a low tactical level and gradually expand to battalion-size combined arms exercises. This would likely involve a habitual Taiwan presence at a U.S. training range, so it would be worthwhile for Taiwan to store relevant training equipment on a long-term basis. For example, U.S. Army units could provide invaluable training in such capabilities as artillery fire and maneuver, UAS operations, the use of IFPC-2 to counter enemy heliborne or UAS operations, and area denial through the use of scatterable mines. The U.S. Army could offer similar train-and-equip assistance to the forces of the Philippines, Vietnam, and other states in the region.

The opportunity to train with U.S. forces would allow Taiwan to develop greater tactical proficiency in some of the critical and very difficult aspects of its evolving defense strategy. It would also help to test whether Taiwan's current concepts are suited to the kind of adversary it is likely to face. Working with the U.S. military on small-unit tactics regarding such capabilities as air defense support to a maneuvering unit, coordination of fire support, and small-unit maneuver under contact could greatly improve Taiwan's proficiency and help to refine its defense concepts.

CHAPTER SIX

The Growing Chinese A2/AD Threat and Blue A2/AD Strategies and Operational Concepts to Counter It

The previous chapters set up the strategic context between China and U.S. allies in the Asia-Pacific region. As noted in Chapter Two, China and U.S. allies in the region have a number of ongoing sovereignty disputes—disputes that would almost unavoidably involve the United States at some level. Also, China's behavior to date has shied away from high-risk military attacks to seize territory, focusing instead on a "salami slicing" approach that leverages diverse economic, diplomatic, and political levers to pressure recalcitrant countries into accommodating Chinese preferences. Chapters Three to Five looked at the potential for such disputes to lead to actual conflicts in Japan, the Philippines, and Taiwan, respectively; at the three allies' strategic and operational concepts to address such conflicts; and at value the U.S. Army could offer its allies in improving its existing strategic and operational concepts.

In this chapter, we focus more broadly on the military threat that China poses in the Asia-Pacific region and then specifically on the threat posed by China's growing A2/AD capabilities. Then we examine how the United States and its allies and partners in the Asia-Pacific region could deter or defeat Chinese aggression protected by its A2/AD capabilities, with a particular focus on Blue A2/AD operations[1]—that is, A2/AD operations conducted by U.S. forces and those of its allies. We next present a number of cases that examine how such operational

---

[1] For an excellent description of Blue A2/AD, see Kelly, Gompert, and Long, 2016.

concepts would work against China. We then briefly discuss Russian A2/AD capabilities and examine a case that involves the Baltics, before closing with another case that deals with the Persian Gulf.

Expected Growth in China's National Power and Its Impact on Its Military Capability

Just as European economies drove growth for much of the previous century, Asia is anticipated to drive most of the world's economic activity by the middle of this century. And much of that economic activity will center on China and the rapidly growing Southeast Asian economic region. Japan, South Korea, and India, however, will remain important drivers of Asia's economic growth and, by extension, the world's economic growth.[2] Even assuming more-pessimistic estimates for China, which put growth at approximately 3–4 percent, and the most optimistic estimates for the United States, which would see growth around 2–3 percent, the size of China's economy will continue to gain relative to the United States.[3]

China's economic growth is, in turn, reflected in the growth of the PLA. The PLA's progress since the mid to late 1990s has been fueled by sustained growth in defense spending, reflecting the high priority that China places on military modernization.[4] Indeed, over the past two decades, rapid economic growth has enabled China to increase its defense budget year after year while managing to maintain defense spending at a relatively low percentage of the nation's GDP. This has allowed Beijing to devote growing resources to national defense with-

---

[2] World Bank, *Global Economic Prospects: Divergences and Risks*, Washington, D.C., 2013.

[3] Office of the Director of National Intelligence, "Global Trends 2030: Alternative Worlds," web page, 2012.

[4] In 2013, China's announced military budget was about $119.5 billion. It is important to account for inflation and other factors, but even when doing so, most analysts believe that China's actual military spending is higher than the announced figures, in part because the official budget excludes some important categories of expenditure, such as procurement of foreign weapon systems and equipment. DoD estimates that China's actual military spending in 2013 was more than $145 billion. Office of the Secretary of Defense, 2014, p. 43.

out shortchanging other important categories of government spending. However, if China experiences a major slowdown in economic growth, there could be sharper trade-offs between defense spending and the government's other budgetary priorities. Additionally, if domestic problems, such as pollution and its associated health costs, continue to worsen, pressure to spend more on these issues could increase, and competition for government budget resources could become more intense.

The trend in the Chinese military's modernization is toward a slightly smaller force capable of fighting wars against advanced militaries. Chinese policy documents call on the PLA to be fully modernized by midcentury, at which point the PLA should be capable of prevailing in wars "under informatized conditions."[5]

Chinese military capabilities have already surpassed those of most of its neighbors, including Taiwan, all Association of Southeast Asian Nations countries, and South Korea. Japan is the only country in Asia capable of fielding forces that are qualitatively superior in comparable numbers; however, most projections foresee China surpassing Japan in terms of numbers of military platforms within a few decades. As the PLA sheds obsolete and outdated equipment in favor of more modern platforms, it will be able to field higher-end airplanes, ships, and missiles in larger numbers than Japan or any other neighboring country does. At some point this century, China's military will very likely stand as the most powerful in numerical terms and could well stand among the most competitive in qualitative terms.[6]

The Chinese A2/AD Threat

China already has the largest naval force in Asia, with 77 principal combatants, 60 submarines, and 55 medium and large amphibious

---

[5] State Council Information Office, *China's National Defense*, white paper, Beijing, 2010.

[6] Michael Swaine, Mike M. Mochizuki, Michael L. Brown, Paul S. Giarra, Douglas H. Paal, Rachel Esplin Odell, Raymond Lu, Oliver Palmer, and Xu Ren, *China and the U.S.-Japanese Alliance in 2030: A Net Assessment*, Washington, D.C.: Carnegie Endowment for International Peace, May 3, 2013.

ships. China is also upgrading its nuclear capability, including a new road-mobile intercontinental ballistic missile—the DF-31A, with a range of 11,200 kilometers—and three Jin-class nuclear-powered ballistic missile submarines (SSBNs) that can carry sea-launched ballistic missiles with a range of 7,400 kilometers. There are at least two more of these SSBNs planned or under production. China's air forces have modernized to a considerable degree as well, with approximately 1,900 aircraft, of which 600 are modern.

China is also improving its ISR capabilities. China has deployed a constellation of BeiDou navigation satellites (NAVSATs) and is expected to complete a global NAVSAT constellation by 2020. China has several remote sensing satellites, which can perform both civil and military applications; a communications satellite; four experimental small satellites; one meteorological satellite; and one manned space mission.[7] China has also invested heavily in an unmanned aerial vehicle (UAV) for long-distance reconnaissance and strike operations. One estimate claimed that China had more than 1,500 UAVs, most of which serve in tactical roles but with a growing number focused on strategic reconnaissance and combat roles.[8]

Chinese training has focused on counterintervention operations involving multiple services. Naval and air forces are operating farther and farther from Chinese shores. The presence of surface action groups operating past the first island chain has grown more common.[9] Chinese forces have also increased joint training focused on small-island seizures.[10]

Within this context, China is rapidly fielding a growing array of capabilities that present a formidable A2/AD challenge. The increasing capabilities that China could bring to bear in the western Pacific have

---

[7] U.S. Department of Defense, *Annual Report to Congress. Military and Security Developments Involving the People's Republic of China*, Washington, D.C., 2014a.

[8] *Defense News*, "Report: China's UAVs Could Challenge Western Domination," June 25, 2013.

[9] U.S. Department of Defense, 2014a.

[10] Matikas Santos, "Chinese Ships Leave Paracel Islands After Landing Drills," *Inquirer*, January 23, 2014.

become a concern to U.S. allies and partners in the Asia-Pacific and to the U.S. military. Chinese A2 capabilities prevent or degrade the ability of forces to enter an operational area, and AD capabilities seek to limit the ability of forces in the operational area to maneuver freely. All potential opponents, whether state or nonstate, have some degree of AD capability.

Chinese A2/AD capabilities include long-range precision theater strike systems—primarily cruise and ballistic missiles and manned aircraft—and anti-ship weapons. These include the DF-11 and DF-15 surface-to-surface ballistic missiles and YJ-83 ASMs; see Table 6.1. China also has one of the largest forces of advanced SAMs, composed primarily of the S-300 and domestically produced HQ-9. These systems can reportedly defend against aircraft and low-flying cruise missiles out to 200 kilometers. In addition, China has fielded multiple rocket launchers with ranges of more than 100 kilometers and extensive cyber and electronic warfare assets.

What Approaches Can the United States and Its Allies and Partners Take to Defeat Aggression Shielded by A2/AD Forces?

There are three broad approaches that the United States and its allies can take to defeating aggression in maritime domains. The first is to *defeat attacking forces, including adversary A2/AD capabilities*. This would require a full-spectrum effort to defeat attacking forces, including (1) long-range ballistic and cruise missiles able to strike an ally's ports and air bases throughout the theater, (2) long-range rocket artillery that can mass fires on targets up to 300 kilometers away, (3) surface ships and submarines that can threaten the deployment of U.S. forces into the region, and (4) cyber and electronic warfare capabilities.

The second approach is to *mitigate the damage caused by enemy attacks*. Avoiding large concentrations of forces, sheltering selected assets, concealing capabilities, and using decoys and other means of deception can mitigate the effects of attacks. Keeping ships and other mobile systems on the move and shifting the location of movable

Table 6.1
Chinese Long-Range Missile Systems

System	Range (km)	Guidance	Type	Munitions	Number of Inventory
WS-1B/A100	180	INS, GPS	Land-attack ballistic missile (potential anti-ship role)	235 kg with up to 500 cluster munitions	Uncertain—but potentially thousands (estimated to cost ~$100K)
WS-2/A200	200	INS, GPS	Land-attack ballistic missile (potential anti-ship role)	235 kg with up to 500 cluster munitions	Uncertain—but potentially thousands (estimated to cost ~$100K)
WS-3/A300	300	INS, GPS, terminal homing guidance	Land-attack ballistic missile (potential anti-ship role)	235 kg with up to 500 cluster munitions	Uncertain—but potentially thousands (estimated to cost ~$100K)
DF-11/A	280–350	INS, GPS, with terminal control	Land-attack ballistic missile (potential anti-ship role)	500–800 kg	700–800
DF-15/A/B	600–800	INS with terminal control	Land-attack ballistic missile (potential anti-ship role)	600 kg	300–400
YJ-83	160	INS, GPS, active/passive radar, infrared	Anti-ship cruise missile	513 kg	Not available

SOURCES: International Institute for Strategic Studies, *The Military Balance 2012*, London, March 7, 2012; Christopher F. Foss and James C. O'Halloran, *IHS Jane's Land Warfare Platforms: Artillery & Air Defence*, London: Jane's Information Group, 2015.
NOTE: GPS = Global Positioning System; INS = inertial navigation system.

assets, such as radars, can increase their survivability. Air bases cannot be moved, but the aircraft can shift among several bases and runway repair capabilities can help airfields recover from attacks.

Finally, the U.S. military and its allies could *impose A2/AD challenges (i.e., Blue A2/AD)* on adversaries attacking over the maritime domain. For example, if the PLA deployed forces by air or sea to seize Taiwan or one of the disputed islands in the South China Sea, the defenders could threaten or negate such a move with the use of anti-ship missiles or air defense systems.

We focus on the potential utility of imposing Blue A2/AD challenges on U.S. adversaries in the following section. The concepts we develop include some measures to mitigate the effects of enemy attacks, as well as a brief discussion of counterbattery fires against adversary A2/AD systems.

Imposing Blue A2/AD Challenges to Deter or Defeat Aggression

As discussed in Chapter One, DoD has voiced its concern that hostile nations might deny "access to the global commons" or conduct A2/AD campaigns.[11] Potential purposes of such A2/AD campaigns might be to secure claims to oil, gas, or other mineral rights; to restrict vessels transiting strategic waterways, especially those carrying military-related supplies or other cargos; or to shield operations to seize territory.

One potential solution is to impose A2/AD challenges—i.e., "Blue A2/AD"—on an adversary's force projection—a solution that would involve increasing the ability of allies and partners with some support from U.S. forces to use ground-based forces to defend territory and interests. While allies and partners might also usefully employ air and naval systems, these forces would face the same challenges that the United States faces in operating aircraft and ships in the presence of highly capable A2/AD systems. Given this, in this approach, we assess

---

[11] U.S. Department of Defense, 2012a.

the potential utility of ground-based capabilities as an asymmetric means of defending contested maritime regions.

A ground-based maritime defense option might work, as shown in Figure 6.1. A hostile nation might use long-range ASMs (capability 1) to keep allied or partner ships out of a contested maritime area. Hostile maritime forces (capability 2) might then sweep the area clear of opposing ships and, depending on their campaign goals, might conduct amphibious operations to seize territory.

A partner nation might oppose these operations by deploying air, sea, or ground sensors (capability 3) to detect, track, and target hostile ships if they attack. If an actual shooting conflict breaks out, partner-nation ASM batteries (capability 4) could engage attacking ships. If an adversary succeeds in landing forces on partner territory—e.g., a

Figure 6.1
Notional Ground-Based ASM Operational Capabilities

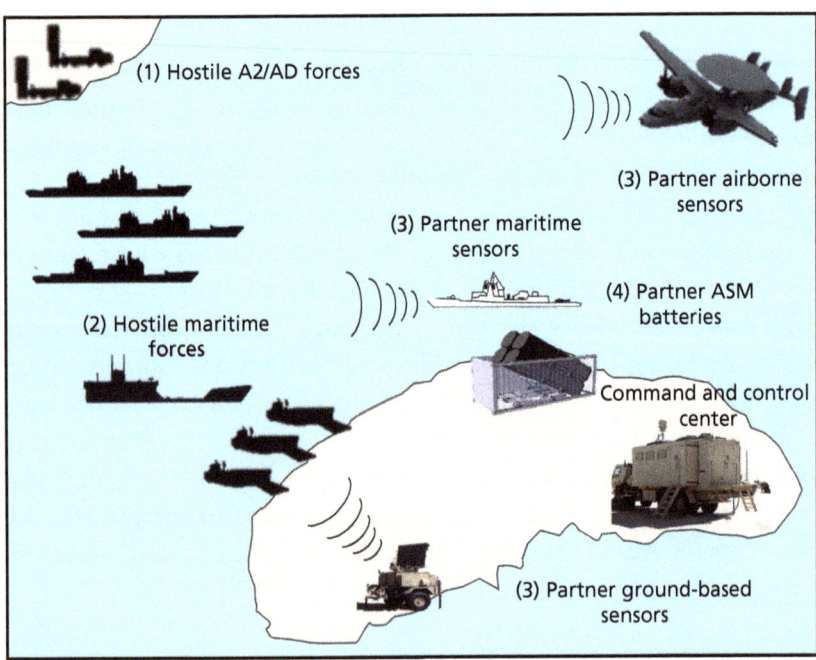

partner island—those forces could be attacked by partner surface-to-surface missile (SSM) batteries based on that island or another island.

The main idea is that allies and partners should be the ones to field a mix of ASM, anti-aircraft, and surface-to-surface fires when they feel sufficiently threatened to employ force in their own defense. Table 6.2 gives several examples of partner and U.S. systems. The primary U.S. role could be to contribute ISR assets and perhaps to assist in providing targeting information and command and control (C2) capabilities. The United States could contribute by employing or providing the systems listed at the bottom of Table 6.2 to complement those from allied and partner nations.

Of course, as shown in Figure 6.2, an aggressive nation would also very likely have other long-range weapons (capability 5) to defeat partner-nation sensors and ASMs. Fighter aircraft and long-range SAMs, such as the S-300, can engage airborne sensors to a range of 200 kilometers. Long-range surface-to-surface rockets, such as the Russian Iskander and the Chinese DF-11, could strike ground-based ASMs at a range of nearly 300 kilometers. And, of course, the aggressors that have long-range ASMs could force ship-based sensors out of range.

A capable adversary would augment its long-range ground-based A2/AD missiles with ships and aircraft. Naval gun and missile fires (capability 6) could help push back or kill sensors on the ground, on ships, or in the air. Such an adversary could also utilize gun and cruise missile fires to attack partner-nation ground-based ASMs. Fighter or bomber aircraft (capability 7) might be even more lethal in attacking partner-nation sensors and ASM batteries. Unopposed, aircraft can loiter for significant periods, finding and attacking partner-nation sensors as they emit and missile batteries as they fire.

In response, partner nations could employ high-altitude airborne sensors and build a distributed ground-based sensor network (capability 8).[12] Each sensor in the network could operate for a period of one

---

[12] For this mission, the U.S. Global Hawk, operating at 65,000 feet, has a horizon of 500 kilometers. This should be sufficient to keep out of range of even the best Russian air defense systems. Fighters, however, can push airborne sensors back—requiring friendly counterair systems to protect airborne ISR assets.

Table 6.2
Exemplar Allied ASM Systems and U.S. Air, Sea, and Ground Missiles

Designation	Range (km)	Guidance	Type	Country of Origin
MM-40 Exocet	70	INS, active radar	ASM	France
Otomat	60–180	INS, data link, active radar	ASM	Italy
Type 12	200	INS, active radar	ASM	Japan
Hsiung Feng III	130	INS, active radar, infrared	ASM	Taiwan
RGM-84	130	INS, active radar	ASM	United States
AGM-84H SLAM-ER	280	INS, GPS, imaging infrared, data link	ASM	United States
Naval Strike Missile	150–200	INS, terrain referential, imaging infrared	ASM	Norway
RBS-15 Mk. III	200–250	INS, GPS, active radar	ASM	Sweden
GMLRS	50–100	GPS/INS	Surface-to-surface ballistic missile	United States
ATACMS	300	GPS/INS	Surface-to-surface ballistic missile	United States
Patriot 2	100	Active radar homing	Anti-aircraft	United States
IFPC-2	5	Imaging infrared	Anti-aircraft	United States

SOURCES: International Institute for Strategic Studies, 2012; Foss and O'Halloran, 2015.
NOTE: SLAM-ER = Standoff Land Attack Missile–Expanded Response; GMLRS = Guided Multiple Launch Rocket System; RBS = Robotsystem.

or a few hours and then swiftly move to a previously prepared location, hide, and get ready to operate again. Preplanned sites in a well-provisioned network would be equipped with fiber optic links to missile batteries and other sensors in the network. While the first sensor is moving, another sensor in the network could take over until it, too,

Figure 6.2
Notional Capabilities to Enhance Survivability of Ground-Based Defenses

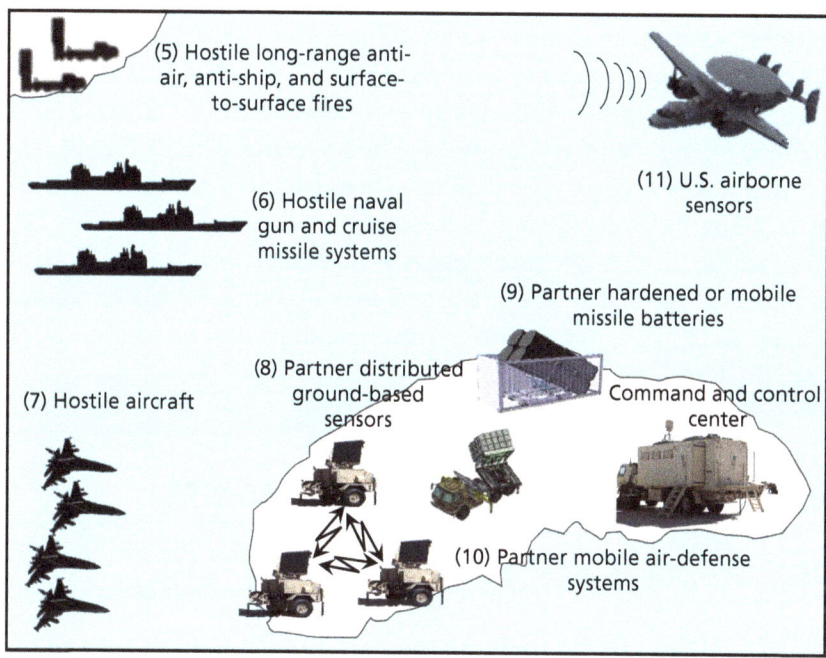

moves. In such a "blinking" network, each sensor moves before the adversary can target the sensor's emissions.

The ASM batteries could similarly survive naval gunfire and cruise and ballistic missiles by employing them on mobile platforms. After firing each volley, the missile battery would move to a new hide site and await orders to fire again. Such "shoot and scoot" tactics made it practically impossible for U.S. aircraft to find and kill Iraqi Scud missiles during Operations Desert Storm and Iraqi Freedom. Alternatively, partner nations could put ASMs in small, hardened bunkers or tunnels (capability 9). The difficulty of precisely targeting and striking such sites might make this a valid alternative, made perhaps more attractive if the firing crews do not need to operate close to the firing systems.

Both the sensors and the ASM batteries could be protected further by mobile air-defense systems (capability 10). A capable ground-based air defense system would raise the costs and risks of hostile aircraft operations in contested areas. These air defenses could be relatively short-ranged systems, such as the IFPC-2, or longer-range systems, such as Patriot. Whichever systems are chosen, it will be crucial that they be mobile to avoid targeting by opposing guns or missiles.

The United States may be able to help partner nations by providing targeting assistance from U.S. airborne platforms (capability 11). These could include such platforms as Global Hawk, if the United States is willing to risk the platforms to long-range SAMs or fighter aircraft. Stealthy manned or unmanned aircraft might be employed for especially high-priority missions.

Imposing Blue A2/AD—Illustrative China Cases

Here we look at three cases—one for Japan, one for the Philippines, and one for Taiwan—that illustrate the notional concepts portrayed in Figures 6.1 and 6.2 involving conflicts with China.

Example Case: Conflict in the Senkakus

Japan would have some geographic advantages if China were to resort to force of arms to pursue its claims over the Senkakus. Japanese air and naval forces based at Yonaguni, Ishigaki, or Iriomote would be within 150 kilometers of the Senkakus, as shown in Figure 6.3. Bases on these islands, however, would be within range of no-notice attacks from DF-15 missiles based in southeastern China. Worse, they are within range of Chinese aircraft operating from the same region of China.

Alternatively, Japan could develop the operational concepts discussed earlier to deter Chinese aggression. The latest Japanese Type 12 missile, with a range of 200 kilometers, can now reach the Senkakus from Yonaguni or Ishigaki. Furthermore, long-range SSMs, such as ATACMS, would threaten any Chinese forces that did manage to land.

Chinese A2/AD Threat and Blue A2/AD Strategies and Operational Concepts   83

Figure 6.3
Maritime Area Coverage from Yonaguni and Iriomote

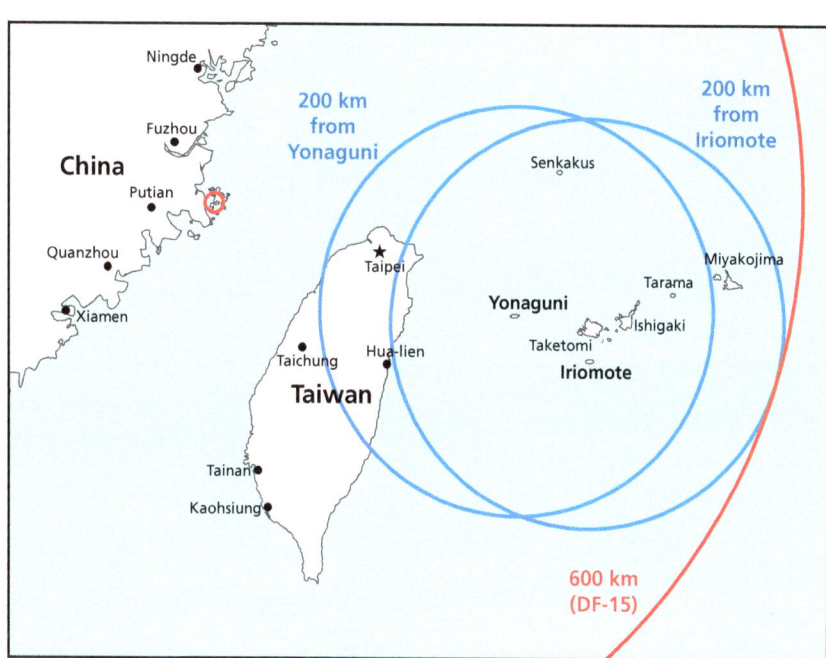

Of the various islands in the southwestern tip of the Ryukyus, Iriomote has some notable advantages for basing A2/AD forces. It is large enough to conceal a significant missile force, has airports and seaports for logistical support, and is close to regional transportation hubs at the neighboring island of Ishigaki. Although Iriomote is sparsely populated, the roughly 2,000 Japanese citizens on the island would be at risk if China were to attack Japanese forces based there (see Figure 6.4).

Japan also plans against the occupation of its uninhabited islands and is equipping and training forces to be inserted onto an uninhabited island.[13] Japan's decision to devote half the GSDF to respond rapidly to such situations points to changes that the country is making to

---

[13] See Ministry of Defense of Japan, 2014, pp. 188–191.

Figure 6.4
Yonaguni and Iriomote Area and Population

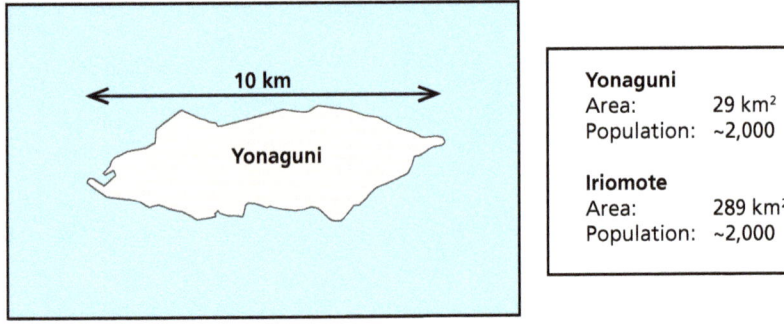

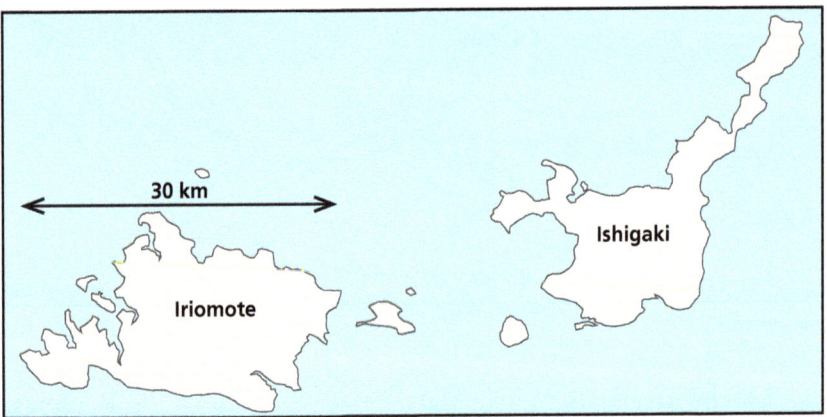

respond to threats to its offshore islands. This capability is clearly relevant when a situation outstrips the capabilities of a police force—for instance, if a sizable group of civilians or a small military force lands on Japanese territory. Japan's defense concept also involves establishing air and maritime superiority in the area surrounding the threatened area.

Japan's concept to introduce ground forces to a conflict over an uninhabited island could be effective if Japan and its adversary both observe some restraint in the conflict. This concept, however, might not be suitable to a conflict that escalates in the Senkakus or other unoccupied islands. In such a situation, it may be difficult for Japan to achieve and maintain the air and maritime superiority that its defensive concept requires to protect the ground forces. In short, the larger the conflict, the more difficult it would be to insert, sustain, or protect GSDF units from heavy air and missile attack.

The firepower that both China and Japan could bring to bear on some of these islands makes a potentially violent conflict especially prone to horizontal escalation. Both countries have multiple ISR and strike systems that give them an ability to deny the other an opportunity to hold disputed territory. Both sides would find it very difficult to supply a force inserted on an uninhabited island and to protect that force from precision attacks. The result could be a stalemate or an expansion of the conflict to other areas, where either side might be able to operate with some advantage.

Example Case: Philippines and the South China Sea

The Philippines is adjacent to the Scarborough Shoal and the Spratly Islands, as well as the waterways surrounding them. The sea areas claimed by China, as demarked by its "nine-dash line," overlap with the internationally recognized Philippine EEZ. The Philippines has an interest in economic activities within its EEZ, and those interests have been upheld by the recent Hague ruling.[14]

China's missile arsenal poses a considerable threat to the Philippines, a threat that is likely to grow in coming years. The map in Figure 6.5 shows the ranges from possible missile launch locations on the Chinese mainland or Hainan Island. The Chinese DF-21C truck-launched ballistic missile[15] and the DH-10 truck-launched cruise missile[16] both have ranges of roughly 1,500 kilometers. They both are very accurate weapons that have circular error probables of less than 100 meters when firing at targets more than 1,500 kilometers from the launch point. Both systems also have a variety of warhead types that could be used to attack fixed facilities, including aircraft parked in the open on airfields. As is clear from the figure, much of the Philippines lies within range of both types of missiles.

---

[14] Permanent Court of Arbitration, 2016.

[15] Sean O'Connor, "PLA Ballistic Missiles," Air Power Australia, January 27, 2014.

[16] Carlo Kopp, "PLA Cruise Missiles," Air Power Australia, January 27, 2014.

Figure 6.5
Ranges from the Chinese Mainland to the Philippines

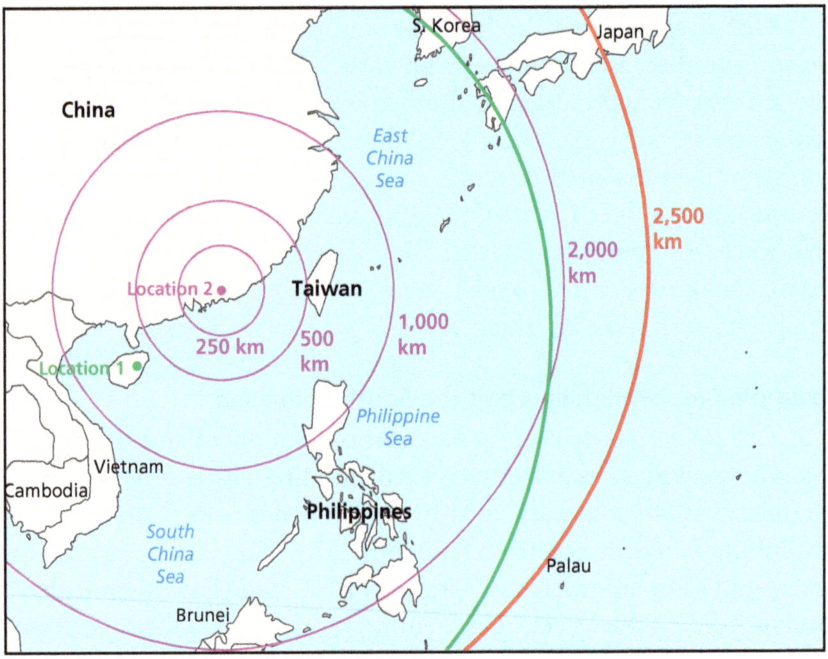

[Map showing concentric range rings from two locations on the Chinese mainland extending to the Philippines, labeled at 250 km, 500 km, 1,000 km, 2,000 km, and 2,500 km. Labels include China, S. Korea, Japan, East China Sea, Taiwan, Philippine Sea, Vietnam, Cambodia, South China Sea, Brunei, Philippines, Palau, Location 1, Location 2.]

RAND RR1820-6.5

To defend its maritime claims, the Philippines would need the ability to control large areas of the South China Sea against well-resourced and growing Chinese naval and air forces. The Philippine Navy, clearly with the PLAN in mind, is purchasing anti-submarine helicopters. If the Philippine Air Force purchases, albeit in limited numbers, either the Swedish Gripen or the U.S. F-16, it will gain a much more effective air defense system than it currently has. Air and naval operations will be particularly challenging while Philippine airports and seaports are under the threat of the missile fires described above.

If the Philippines decides that it must be able to defend its claims, then land-based missiles could offer a cost-effective way to do so. There are three types of land-based fires that might be applicable in the case of the Philippines: (1) air and missile defense, (2) anti-ship weapons, and (3) long-range surface-to-surface strike systems. The Philippine military is now moving to create capabilities of the first two types, but how extensive and in what time frame remain uncertain. Each of these possibilities will be discussed below from both a U.S. Army and Philippine military perspective.

The Philippine military is acquiring an ASM capability to be deployed by the Philippine Army. It is important to the United States to understand how and under what conditions the Filipinos would employ these systems. The Philippines' main territorial dispute with China is now over the Spratly Islands in the South China Sea. The Philippine government might elect to deploy coast-defense missiles as one way of asserting its sovereignty over the Spratlys. The Philippine home islands could offer large operating areas for mobile missiles with ample opportunities to conceal them from attack (see Figure 6.6). Missiles with a 300-kilometer range could just cover the Scarborough Shoal; however, ASMs would need 370-kilometer ranges to cover the entire Philippine EEZ.

Figure 6.6
Coverage of the South China Sea from the Philippines

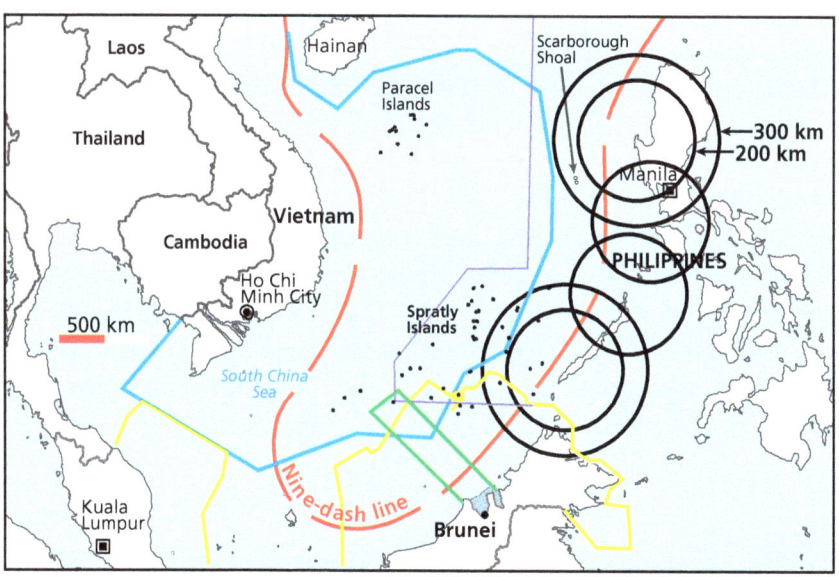

The Philippines might not see a need for ASMs to defend the main Philippine islands from Chinese naval attack, blockade, or invasion. And the Philippines might view the continued operations of the U.S. Navy and U.S. Air Force in the South China Sea as sufficient to counter an attack by the surface elements of the PLAN. Also, even if U.S. air and surface naval elements were forced to keep their distance from the Chinese mainland because of the PLA's A2 capabilities, U.S. Navy submarines operating in the South China Sea would still threaten Chinese warships operating hundreds of kilometers from their ports.

The Philippine armed forces have no meaningful missile defenses and are currently in discussion with Israel for a still-to-be-determined number of systems. These new systems could be entirely focused on aircraft and cruise missiles, or they could be intended for ballistic missile defense. Regardless of what combination of Israeli-designed systems the Philippine military elects to purchase, the cost of air and missile defenses will be considerable. It is likely that the Filipinos will only be able to deploy a limited capability to defend very high-value military, economic, and political assets.[17]

Thus far, the Philippine military has not expressed an interest in long-range, land-based strike missiles. The Scarborough Shoal is less than 250 kilometers from central Luzon island, putting the shoal within range of ATACMS-class rocket artillery. Such artillery could effectively deny China use of this and other artificial islands within range of the Philippine home islands.

Example Case: Taiwan

The island of Taiwan, with a total area of 36,000 square kilometers and separated by 200 kilometers from the mainland, is well suited to defensive operations (see Figure 6.7). The western third of the island features many cities and towns with a well-developed road network. The central and eastern two-thirds of the island features forested or mountainous terrain.

---

[17] *ID James*, "Israeli Arms Sales to Philippines and Its Implications: The Present Scenario," August 2013.

Figure 6.7
Coverage of Maritime Approaches to Taiwan

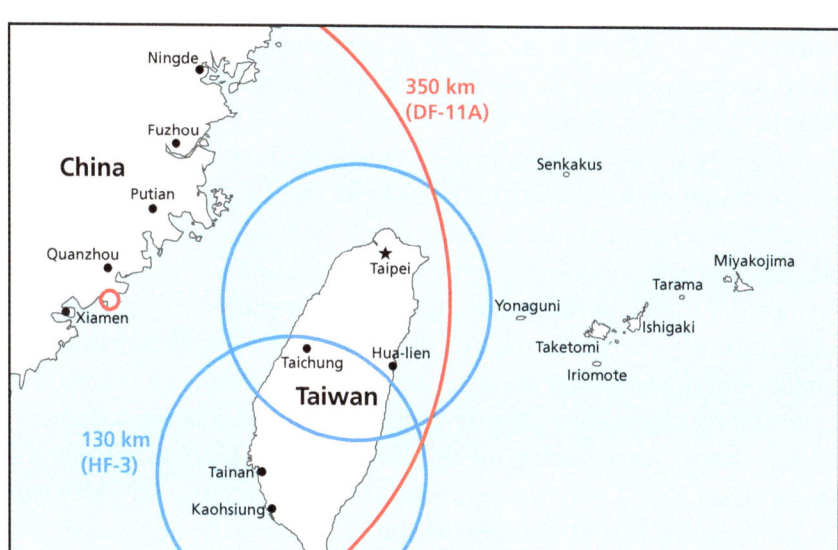

In particular, Taiwanese geography is conducive to the missile-based defensive concepts described earlier in this chapter. Taiwan has more than 40,000 kilometers of roadways for mobile systems to operate and move along. Its natural geography could provide plentiful locations for prepared hiding sites for missile batteries. Finally, Taiwan's ASMs—with a range of 130 kilometers—could cover its coast and approaches from a handful of firing locations. It is notable, however, that all of Taiwan can be hit with Chinese missiles with ranges of 350 kilometers. The Chinese DF-11A, for example, could hold all Taiwan at risk from just one firing location.

Countering a PLA amphibious assault is obviously a high-priority objective for Taiwan. Analysis suggests that Taiwan's armed forces could, at fairly modest expense, field much more-effective self-defense capabilities against this type of threat than they currently have. Taiwan has both the economic means and the technical and operational capacity to develop, deploy, and operate systems, such as short-range UASs and anti-ship cruise missiles, shallow water mines, rocket artillery, mobile short-range air defenses, and communications jamming gear—all of which, properly employed, could significantly contribute to an effective defense against invasion.[18]

As noted, Taiwan has already made investments in wide-area air defense systems with the Patriot PAC-3 and the TK-III systems. While Patriot PAC-3 systems have some capabilities against ballistic missiles, the systems are unlikely to be the best use of these capabilities for Taiwan. All missile defenses have limits of how many missiles they can simultaneously engage. In the case of China, which could launch many missiles simultaneously, even a perfectly effective defense is susceptible to saturation. In such a situation, using a missile defense system to defend a fixed target, such as an air base, plays into an attacker's hands. The attacker will know the location of such defenses and then can choose to try and avoid it or overwhelm it. Either way, the attacker has the advantage.

Worse, long-range rocket artillery, such as the Chinese WS-1B, WS-2, and WS-3, can fire across the Taiwan Strait. For example, the Chinese WS-2 is a guided artillery rocket with a per-rocket cost of approximately $100,000. The WS-2 multiple rocket launcher fires 400-millimeter rockets with ranges of roughly 200 kilometers, giving it the ability to fire across the Taiwan Strait. Future versions of the WS-2 are projected to have ranges of 350 kilometers.[19] Patriot is capable of engaging the WS-2 and other cruise missiles and rockets, but it is an expensive system. The current price for the Patriot PAC-3 is roughly $4 million per missile.[20] There would be many advantages if

---

[18] See Michael Lostumbo, "A New Taiwan Strategy to Adapt to PLA Precision Strike Capabilities," in Roger Cliff, Phillip C. Saunders, and Scott Warren Harold, eds., *New Opportunities and Challenges for Taiwan's Security*, Santa Monica, Calif.: RAND Corporation, CF-279-OSD, 2011.

[19] *Military-Today*, "WS-2: Multiple Launch Rocket System," web page, undated.

[20] Joakim Kasper Oestergaard Balle, "MIM-104F Patriot PAC-3: About Patriot and PAC-3," BGA-Aeroweb, September 6, 2016.

instead Taiwan used the Patriot to defend against aircraft, something for which it is extremely well suited. Taiwan can operate these defenses at a time of its choosing and to support important operations—for instance, a counterattack by the Taiwanese Army.

The second consideration is that, while Taiwan now has very good wide-area and long-range air defense coverage from its Patriot and TK systems, its shorter-range systems are antiquated. Investments in more-modern short-range defenses would complement the current ground-based air defense systems.[21]

Taiwan already operates land-based, mobile, anti-ship, cruise missile capabilities. The key to their effectiveness is surviving long enough to launch their missiles and receive precise cues to target the missiles in the presence of ocean clutter and passive and active countermeasures. The survivability of the launchers will come down to the discipline of the crew to keep these truck-mounted missiles hidden until they are needed and then to move to a suitable launch site and quickly launch and hide to avoid presenting a lucrative target. The end-game effectiveness of the missile is going to be determined by the improvements made to the seeker.

The Russian A2/AD Threat

Our focus in this report is on China, but Russia also poses an A2/AD threat. To evaluate fully the potential utility of Blue A2/AD strategies, we must extend our analyses beyond the Asia-Pacific region and into other theaters of potential conflict. Here, we briefly discuss the Russian A2/AD threat and offer two cases. After those two cases, we provide a case in the Persian Gulf.

The vulnerability of the Baltic nations to a short-notice Russian attack is aggravated by Russian militarization of the strategically located Kaliningrad enclave. Long-range Russian missiles based in Kaliningrad, such as the SSN-26, with a range of approximately 300 kilometers, could block NATO vessels from entering the Baltic Sea. Together

_____

[21] Lostumbo et al., 2016.

with other well-placed sites in western Russia, Russian missiles could attack shipping into all the Baltic nations and Poland.

Paired with long-range S-300 and S-400 SAMs, the Russians could deny NATO sea and air forces access to the Baltics. The Russian air force and Baltic fleet could then operate with relative impunity—for example, to support attacks against Estonia and Latvia; mount amphibious landings in their rear areas; or take other strategic areas, such as the Swedish island of Gotland. The S-300 can operate at ranges of up to 200 kilometers, and the S-400 is reputed to be effective at ranges of up to 350 kilometers (see Table 6.3). Russian surface-to-air and antiship operations could also be supported by the Iskander surface-to-surface ballistic missile as a counterbattery system or in an anti-ship role if it is equipped with a terminal guidance package able to target ships.

Example Case: Baltic Sea Control

Recent Russian aggression in Ukraine has raised concerns that NATO's eastern flank is vulnerable to attack. These fears have been heightened by Russian security strategy that describes the "NATO infrastructure" bordering Russia as a threat and by the large and frequent Russian "snap exercises" of its armed forces. The Baltic nations of Estonia, Latvia, and Lithuania are particularly vulnerable to Russian aggression given their proximity to Russian territory (see Figure 6.8).

To strengthen NATO's ability to control maritime operations in the Baltic Sea, NATO might consider augmenting its naval and

Table 6.3
Russian Long-Range Missile Systems

System	Range (km)	Guidance	Type	Country of Origin
SS-N-26	300	INS, active and passive radar	ASM	Russia
S-300	200	INS, active and passive radar	Anti-aircraft	Russia
S-400	350	INS, active and passive radar	Anti-aircraft	Russia
Iskander	280	INS, GPS, optical homing	SSM	Russia

SOURCES: Foss and O'Halloran, 2015; Malcolm Fuller and David Ewing, *IHS Jane's Weapons: Naval*, London, Jane's Information Group, 2013.

air forces with long-range ASMs. Using systems with a range of 200 kilometers—such as Norway's Naval Strike Missile or Sweden's Robotsystem 15 (RBS-15) (see Table 6.2)—would allow NATO to cover nearly the entire Baltic Sea from as few as three locations, as shown in Figure 6.8. (As we will discuss in Chapter Seven, Poland has begun to acquire the Naval Strike Missile from Norway and could be the lead nation for deploying these weapons.) This would help to deny Russia the ability to mount amphibious operations or to use surface ships to interrupt sea traffic making runs from Stockholm to ports in Estonia or Latvia. Such missiles would also enable Sweden to extend the defense of its own coastline, Gotland, and other strategic islands well into the Baltic Sea. Interestingly, Sweden has announced that it will reactivate its land-based ASM defenses.[22]

It is also interesting to consider the potential value of long-range surface-to-surface fires in the Baltics. Consider the ability of ATACMS-class missiles, with a range of 300 kilometers, in Figure 6.9. From

Figure 6.8
Coverage of the Baltic Sea from Russia and NATO Nations

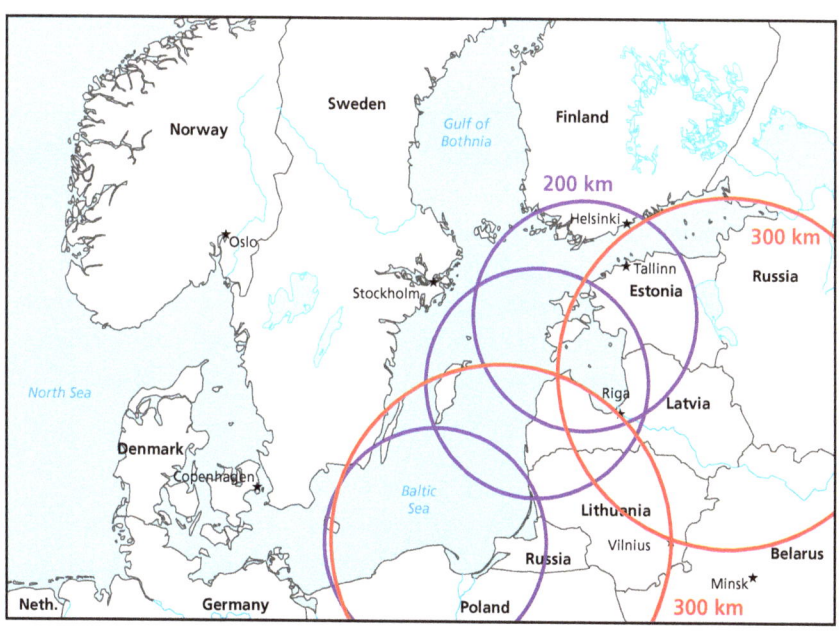

Figure 6.9
Coverage of the Baltics, Belarus, and Russia from Poland and Estonian Coastal Islands

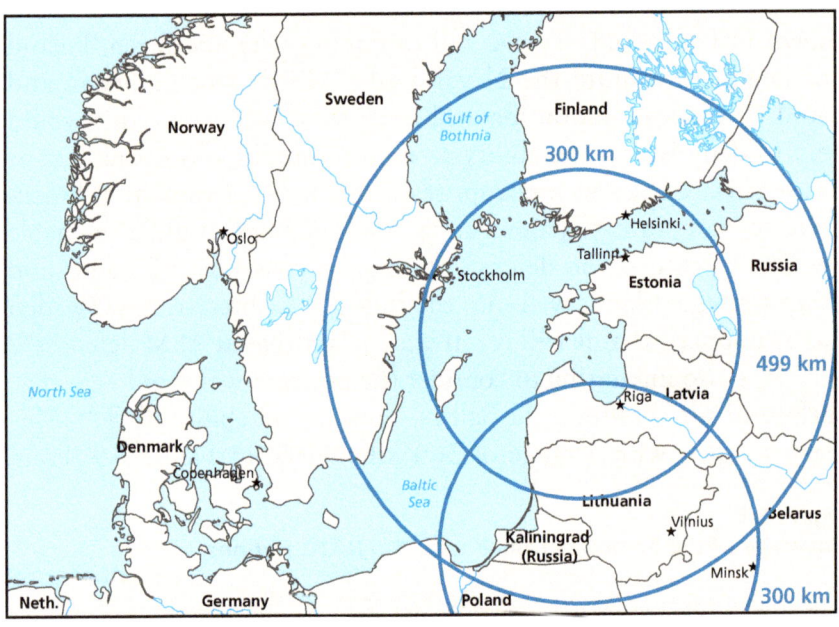

just two launch points, one in northeastern Poland and the other on one of the islands off the coast of Estonia, NATO forces could cover nearly the whole of the Baltic states, much of western Belarus, and all of Kaliningrad with suppression of enemy air defenses (SEAD) or counterbattery fires. Improving the range of NATO missiles to the Intermediate-Range Nuclear Forces (INF) Treaty–compliant range limitation of 499 kilometers would extend SEAD and counterbattery coverage into western Russia from a single site, notionally placed on the islands off the coast of Estonia.

Example Case: Defense of Black Sea NATO Allies

NATO might also wish to expand its options to defend the Black Sea coasts of Romania, Bulgaria, and Turkey. Missiles with a 200-kilometer range could cover the whole of the Romanian and Bulgarian coasts, as well as the Bosphorus Strait (next to Istanbul), from as few as two sites (see Figure 6.10). Deploying these missiles to bases along the length of Turkey would extend NATO ASM coverage over the whole of the southern Black Sea.

Defending the Persian Gulf

Although Iran poses a small maritime threat to U.S. partners in the Persian Gulf, such nations as Kuwait, Bahrain, Saudi Arabia, Qatar, and the United Arab Emirates might find ground-based surface-to-surface and anti-ship missiles to be useful in their maritime defense concepts. Together, the Gulf Cooperation Council nations would be able to cover the whole of the Persian Gulf against potential Iranian naval provocations or harassment with just a few systems based at key places. Iranian provocations might include emplacing mines, harass-

Figure 6.10
Coverage of the Black Sea and Bosphorus Strait from NATO Nations

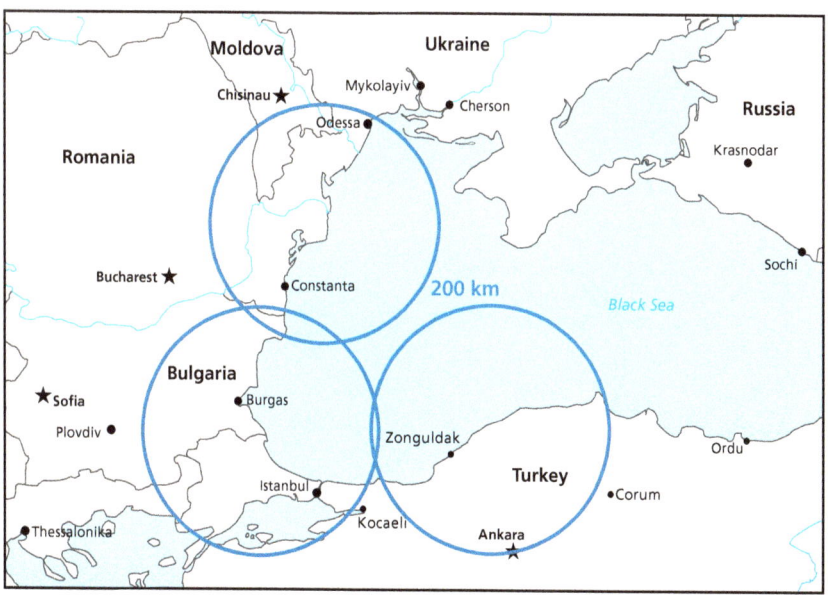

RAND RR182-6.10

ing or boarding civilian vessels, or threatening the naval forces of U.S. partners as they patrol the gulf.

For example, a missile with a 200-kilometer range could cover the whole of the Persian Gulf, including the Strait of Hormuz, from three carefully chosen sites (see Figure 6.11). Such coverage would help ensure that any Iranian naval actions could be countered from ground bases and air and naval platforms.

Alternatively, the Iranians could utilize ground-based ASMs to attack ships passing through the Strait of Hormuz. Gulf Cooperation Council nations might be able to target Iranian A2/AD systems attempting to close the strait if they possess rocket artillery with sufficient range.

Iranian ground-based launchers might be vulnerable to counter-battery fire from long-range rocket artillery systems with the range of

Figure 6.11
Coverage of the Persian Gulf

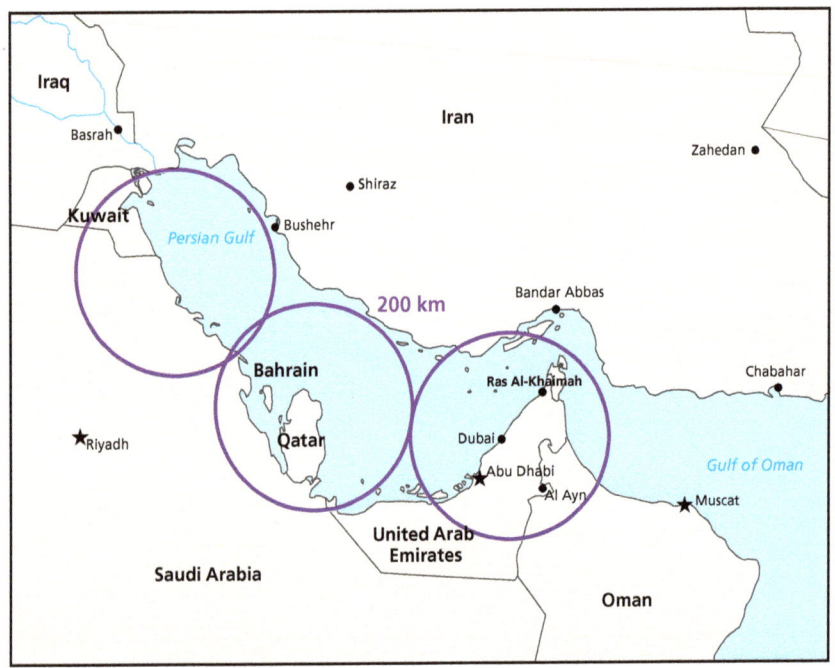

RAND RR1820-6.11

the ATACMS class. These same long-range batteries might be useful to suppress Iranian ballistic missile batteries or surface-air missile batteries. For example, from one launch point at the tip of the United Arab Emirates, an ATACMS-class missile with a 300-kilometer range could conduct SEAD or counterbattery fires against the entire Iranian side of the Strait of Hormuz (see Figure 6.12).

Figure 6.12
Coverage of Iranian Bases from the Arabian Peninsula

CHAPTER SEVEN
Potential Roles for U.S. Land-Based Fires in Joint Missions

In the previous chapter, we considered the role land-based fires might play in assisting U.S. allies and partners in providing for their own defense. In this chapter, we assess the role land-based systems might play in controlling or denying hostile operations on the sea, in the air, or on the ground as part of the operations of U.S. joint forces. In general, we focus on U.S. actions intended to aid U.S. allies and partners, but we also consider cases in which the United States might act alone to counter aggression. We examine constraints on U.S. land-based fires and the potential U.S. Army contributions from land-based fires, and we estimate the costs to field land-based missile systems. The focus here is on the PACOM AOR, but we also discuss the use of land-based fires in dealing with Russia.

Constraints on U.S. Land-Based Fires

Employing U.S. Army fires in the Pacific requires resolving an obvious dilemma: identifying land from which to employ the land-based fires systems. The PACOM AOR features vast expanse–s of sea separating islands and landmasses. Figure 7.1 depicts the western Pacific portion of the PACOM AOR and highlights representative locations where land-based fires could either be based or be targeted against.

Geographic constraints combine with other important policy factors to constrain the role of U.S. land-based fires. The weapon

**Figure 7.1
Map of the PACOM AOR**

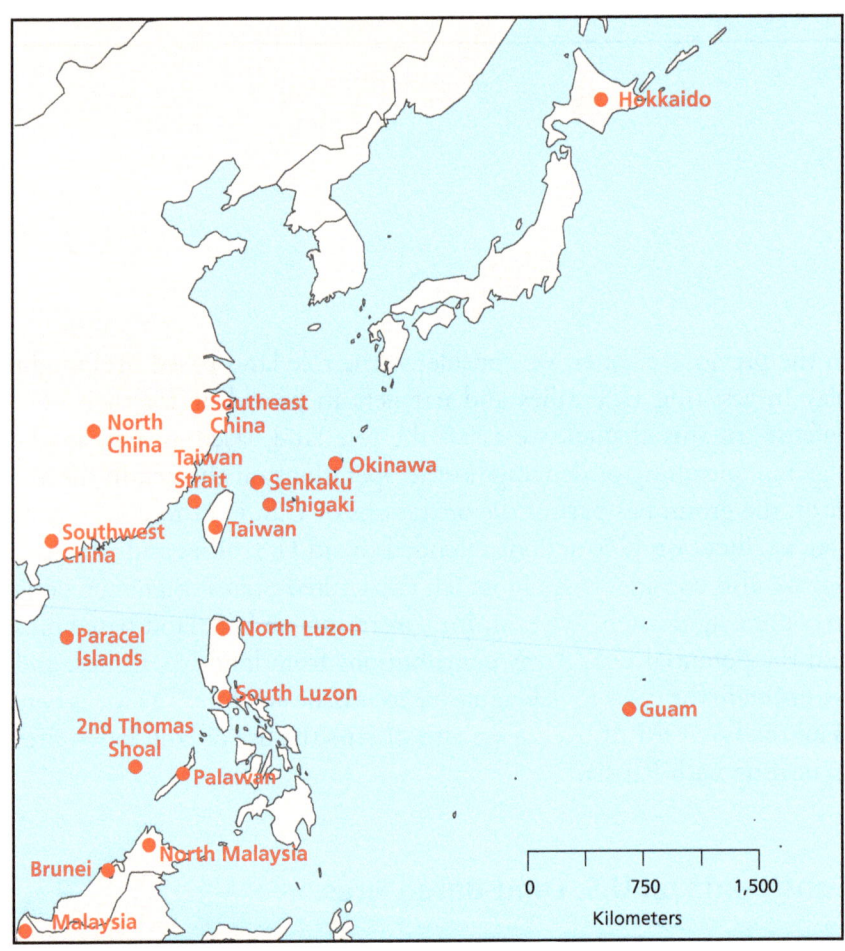

with the longest range in the Army's inventory is ATACMS, with a reach of roughly 300 kilometers. This range will be improved when the Long-Range Precision Fires (LRPF) missile—with a range of 499 kilometers—is delivered. That missile, however, is not scheduled for delivery until 2027. The United States is a signatory to the INF Treaty, which precludes the United States from having land-based cruise and ballistic missiles with ranges of 500 to 5,500 kilometers. To put these ranges into perspective, even missiles based in the northern Philippines (Luzon) and Okinawa are farther than 500 kilometers from the Taiwan Strait.

Moreover, the United States has not taken sides in the disputes that China has with Japan over the Senkakus, with the Philippines over the Spratlys, or with the reunification of Taiwan, other than to insist that such disputes be resolved peacefully. This policy limits where U.S. forces could be prepositioned in anticipation of a crisis. And, of course, there are political questions about whether partners would allow the United States to execute offensive fires from their territories, before, during, or after a crisis erupts—particularly if the partner in question is not otherwise directly involved in the conflict. This is where the advantages of naval forces are most apparent, because they have the ability to hold adversaries at risk from nearby international waters. To a lesser degree, the Air Force faces the same limitations as the Army in employing force from one allied nation to assist in the defense of another, although the Air Force can bridge the distance from bases to targets by using tankers and standoff weapons.

Finally, there are infrastructure limitations, because many small islands in the western Pacific lack airports or seaports from which to receive forces or are home to rugged terrain or dense jungles that would inhibit basing and operations.

Table 7.1 illustrates the size of the AOR by summarizing the distances between the operationally and strategically representative locations shown in Figure 7.1. The distances are vast. Even where international agreements make possible the use of land-based fires, the distances involved make it difficult for the United States to use bases in one allied nation to support operations to defend another.

The effects of geographic, policy, arms control, and current weapon-system limitations are depicted in Table 7.2, which mirrors the structure of Table 7.1. A cell is colored red if multiple constraints preclude employing Army fires from a base (row) or to a location (column); yellow if only one factor constrains employment; and green if there are no fundamental barriers. Each cell is annotated with the limiting factors.

Table 7.1
Approximate Distances Between Representative Locations in the PACOM AOR (km)

Base/Target	Taiwan Strait (Ships)	Taiwan	Senkakus	Paracel Islands	Spratly Islands (2nd Thomas Shoal)	SE China	SW China	N China
United States (Hawaii)	8,280	8,210	7,890	9,320	9,230	8,120	9,170	8,740
United States (Guam)	2,890	2,760	2,600	3,500	3,170	3,100	3,650	3,580
Taiwan	150	0	350	1,190	1,650	630	970	850
Japan (Hokkaido)	2,940	2,960	2,610	4,130	4,550	2,550	3,690	3,080
Japan (Okinawa)	850	780	450	1,960	2,270	850	1,730	1,410
Japan (Ishigaki)	440	340	170	1,500	1,850	690	1,310	1,100
Japan (Senkakus)	390	350	0	1,550	1,960	530	1,280	1,000
Philippines (North Luzon)	700	590	850	950	1,120	1,220	1,090	1,300
Philippines (South Luzon)	1,170	1,060	1,310	980	760	1,690	1,370	1,710
Philippines (Palawan)	1,680	1,590	1,870	1,030	290	2,200	1,610	2,110
Spratly Islands (2nd Thomas Shoal)	1,710	1,650	1,960	860	0	2,220	1,490	2,050
Spratly Islands (Itu Aba)	1,690	1,640	1,970	730	180	2,190	1,380	1,970
Brunei	2,310	2,240	2,550	1,370	600	2,820	2,030	2,620
Malaysia	2,510	2,450	2,760	1,520	800	3,020	2,170	2,790
North Indonesia	2,360	2,290	2,580	1,500	680	2,880	2,150	2,720
North Malaysia	2,080	2,000	2,290	1,260	410	2,590	1,900	2,450

Table 7.2
Net Effect of Geography, Arms Control Agreements, Policy, and Current Weapon Systems

Base/ Target	Taiwan Strait (Ships)	Taiwan	Senkakus	Paracel Islands	Spratly Islands (2nd Thomas Shoal)	SE China	SW China	N China
United States (Hawaii)	INF Treaty; limitations of current weapon systems							
United States (Guam)	INF Treaty; limitations of current weapon systems							
Taiwan	Policy precludes pre-positioning U.S. forces; limitations of current weapon systems	Policy precludes prepositioning U.S. forces		Policy precludes prepositioning U.S. forces; INF Treaty; limitations of current weapon systems				
Japan (Hokkaido)	INF Treaty; limitations of current weapon systems							
Japan (Okinawa)	INF Treaty; limitations of current weapon systems		Limitations of current weapon systems	INF Treaty; limitations of current weapon systems				
Japan (Ishigaki)	Limitations of current weapon systems			INF Treaty; limitations of current weapon systems				
Japan (Senkakus)	Policy precludes pre-positioning U.S. forces; limitations of current weapon systems	Policy precludes prepositioning U.S. forces		INF Treaty; limitations of current weapon systems				
Philippines (North Luzon)	INF Treaty; limitations of current weapon systems							

Table 7.2—Continued

Base/ Target	Taiwan Strait (Ships)	Taiwan	Senkakus	Paracel Islands	Spratly Islands (2nd Thomas Shoal)	SE China	SW China	N China
Philippines (South Luzon)	INF Treaty; limitations of current weapon systems							
Philippines (Palawan)	INF Treaty; limitations of current weapon systems						INF Treaty; limitations of current weapon systems	
Spratly Islands (2nd Thomas Shoal)	Policy precludes prepositioning U.S. forces; INF Treaty; limitations of current weapon systems				Policy precludes prepositioning U.S. forces	Policy precludes prepositioning U.S. forces; INF Treaty; limitations of current weapon systems		
Spratly Islands (Itu Aba)	Policy precludes prepositioning U.S. forces; INF Treaty; limitations of current weapon systems				Policy precludes prepositioning U.S. forces	Policy precludes prepositioning U.S. forces; INF Treaty; limitations of current weapon systems		
Brunei	INF Treaty; limitations of current weapon systems							
Malaysia	INF Treaty; limitations of current weapon systems							
North Indonesia	INF Treaty; limitations of current weapon systems							
North Malaysia	INF Treaty; limitations of current weapon systems				Limitations of current weapon systems	INF Treaty; limitations of current weapon systems		

NOTE: A cell is colored red if multiple constraints preclude employing Army fires from a base (row) or to a location (column); yellow if only one factor constrains employment; and green if there are no fundamental barriers.

We see that only a few of the base-target pairs in this sampling present an opportunity: deployments at Ishigaki for a crisis over Taiwan or the Senkakus and deployments on Palawan for some targets in the South China Sea. But these pairs, too, may be bound by political restrictions and infrastructure requirements limiting their use. For example, it is not certain that Japan would allow U.S. strike platforms on Ishigaki in a campaign to defend Taiwan. The distance from Palawan to the Second Thomas Shoal would require systems with a range of 300 kilometers to be effective, perhaps including ATACMS operations against Chinese bases on reclaimed reefs or long-range ASMs operating in proximity of the reefs.

While facing the same geography, potential adversaries might not be operating under the same policy constraints. For example, Figure 7.2 depicts the areas that China can cover with ballistic and cruise missiles currently in its inventory. A nation not bound by the INF Treaty, China has built systems that can reach virtually all the potential U.S. and allied land bases noted above.

Later, we discuss that the United States and its NATO allies in Europe face shorter distances and fewer formal policy constraints.

Figure 7.2
Chinese Missile Coverage of Western Pacific

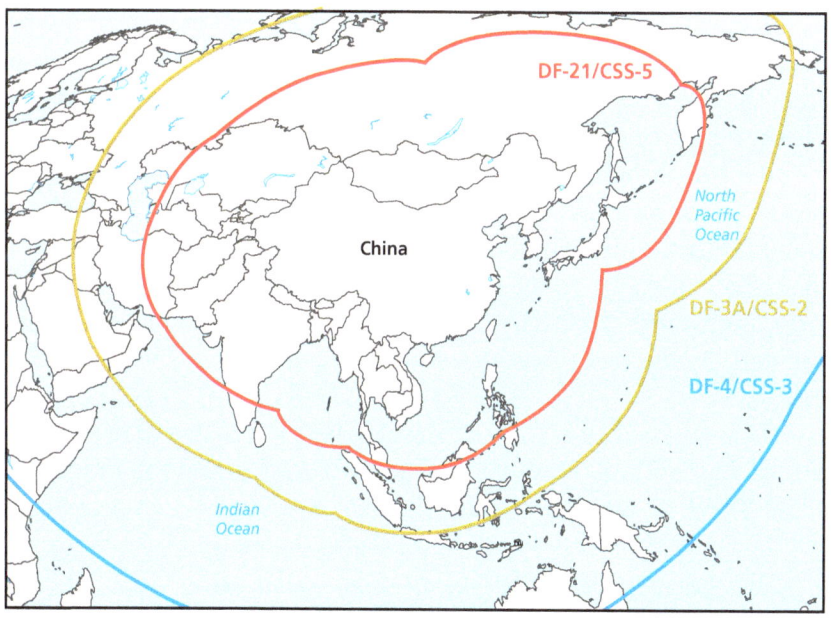

Potential Army Roles in Anti-Ship Operations

In this section, we examine the demands and potential roles for the U.S. Army in anti-ship operations. We consider anti-ship missions in three categories: (1) close-in tactical coast defense against an amphibious assault (2) longer-range maritime interdiction against military vessels, and (3) lethal enforcement of a blockade.

Close-in Tactical Coast Defense Against an Amphibious Assault

Close-in tactical coast defense is focused on defending against nearby enemy naval vessels and landing craft approaching a beach. Weapons used for this purpose typically have a relatively limited range—100 kilometers or less in most cases. Often these weapons engage naval targets that are within sight of land, reducing the need for long-range targeting. An example is the truck-launched Norwegian Penguin ASM that has a range of roughly 50 kilometers and delivers a 120-kilogram warhead to the target. This type of weapon would obviously be useful against approaching landing craft that are attempting an amphibious assault.

If there is a regional war in the Pacific, for example, the Chinese could attempt to seize disputed territories through an amphibious invasion. The canonical case tends to be a large-scale invasion of Taiwan by China, but the Chinese could also seize reefs in the South China Sea; the smaller, offshore islands controlled by Taiwan; or the Senkakus. Similarly, the Swedish island of Gotland and the Estonian islands of Saaremaa and Hiiumaa occupy strategic positions in the Baltic Sea and, thus, might be targeted by Russia if it attacks NATO. In these cases, short-range, land-based ASMs could be useful to deter an invasion and, if deterrence fails, defeat it, as described in previous RAND analyses.[1]

U.S. regional allies and partners are best positioned to deploy the needed systems and, presumably, have the most to gain by procuring them. The most practical role for the U.S. Army may be to help build

---

[1] David A. Shlapak, David T. Orletsky, Toy I. Reid, Murray Scot Tanner, and Barry Wilson, *A Question of Balance Political Context and Military Aspects of the China-Taiwan Dispute*, Santa Monica, Calif.: RAND Corporation, MG-888-SRF, 2009.

partners' indigenous coastal defense capabilities, as we described in the previous chapter. The U.S. Army might help allies and partners build the operational concepts to effectively use ASMs and the units, force structure, and institutions to build and maintain a ready capability. Allies and partners that already have some anti-ship capabilities, such as Japan and Poland, might benefit by establishing common standards with the U.S. Army—e.g., for receiving ISR and targeting information, preventing fratricide, continuing to develop new concepts, and engaging in combined training exercises.[2] This approach would present opportunities for the Army to help build partner capability, but it would not necessarily require a large U.S. investment in new systems, technology, or force structure.

The U.S. military could also provide some of the ISR capabilities to provide tactical warning and the targeting information needed to employ short-range ASMs effectively. Most of this information would likely be provided by sensors mounted on U.S. aircraft and ships; however, ground-based sensors might also be useful in some situations.

Finally, the U.S. Army might reinforce allies and partners if they were to engage in defensive combat operations against an attacking force. Several factors, however, presently limit the ability of the U.S. Army to execute this mission. First, the U.S. Army does not currently have purpose-built systems to conduct tactical coastal defense. These systems would have to be acquired, and units would need to be identified and trained to employ them. Second, actually employing these units would require taking sides in a regional dispute—e.g., by putting U.S. troops on Taiwan or in the Ryukyus. This could be a destabilizing act and could prompt an adversary to strike first or to wait and strike after U.S. forces are eventually withdrawn.

If deterrence fails, the United States would find it hard to deploy land-based anti-ship capabilities once the shooting starts, which is why it would be preferable for Taiwan to field such systems. Similarly, in a

---

[2] Benjamin Schreer, "China's Growing Military Might Has Japan on Edge: Tokyo Responds," *The National Interest, The Buzz*, August 8, 2014; Zachary Keck, "Taiwan Acquires Submarine-Launched Anti-Ship Missiles," *The Diplomat*, December 27, 2013.

South China Sea scenario, U.S. allies could use such systems to hold Chinese forces at risk.

Long-Range Maritime Interdiction Against Military Vessels

A second potential anti-ship mission is long-range maritime interdiction against attacking vessels. The relevant weapons are typically longer-range and can be air-, ground-, or ship-launched systems. An example is the Chinese C-803 ground-launched missile that has a range of roughly 350 kilometers or the Norwegian Naval Strike Missile with a range of 200 kilometers. Although these weapons could be used against close-in targets in the same way as the shorter-range systems mentioned above, their main role is to interdict naval targets at much longer ranges.

Maritime interdiction missions could be useful in a wide range of scenarios, including Chinese aggression against the Senkakus, seizure of reefs in the South China Sea, or an invasion of Taiwan. In fact, maritime interdiction could be preferred to coastal defense if such missions could be conducted at standoff distances away from Chinese A2 capabilities.

The United States has the capability to conduct maritime interdiction missions, represented by U.S. attack submarines equipped with Mark 48 torpedoes and carrier air wings with Harpoon missiles, and DoD is executing plans to improve those capabilities. Although the Air Force has divested the capability to conduct anti-ship missions using standoff weapons, the Air Force and Navy are collaborating to develop a new ASM that could be launched from bombers at standoff ranges of 500 miles or more.[3] Table 7.3 summarizes the current and planned anti-ship capacity. For comparison, Table 7.4 shows the current and projected size of the Chinese Navy.[4]

There is a question about whether this capacity, and the capability it provides, could be brought to bear at the specific times and in the places where it might be needed. If land-based anti-ship forces were needed to complement existing joint force capabilities, it would

---

[3] Schreer, 2014.

[4] Heginbotham et al., 2015.

Potential Roles for U.S. Land-Based Fires in Joint Missions   109

Table 7.3
Existing Joint Anti-Ship Delivery Capacity

Joint Anti-Ship Capability	Current Inventory (2015)	Planned Inventory (2020)	Planned Inventory (2025)	Notional PACOM Crisis Deployment	Aggregate Anti-Ship Launch Capacity
Attack submarine	58	53	51	10–20	240–520 Mark 48 torpedoes[a]
Carrier air wing	10	11	11	3–5	3–5 carrier air wings
Bomber	157	159	159 (fiscal year 2023)	30–60	480–1,440 JASSM-sized weapons

SOURCES: Office of the Chief of Naval Operations, *Report to Congress on the Annual Long-Range Plan for Construction of Naval Vessels for FY2015*, Washington, D.C.: U.S. Department of Defense, June 2014; U.S. Department of Defense, *Annual Aviation Inventory and Funding Plan: Fiscal Years (FY) 2014–2043*, Washington, D.C., May 2013.

NOTE: JASSM = Joint Air-to-Surface Standoff Missile.

[a] Up to 24 torpedoes on a Virginia-class submarine and up 26 on an improved Los Angeles–class submarine (Periscope).

Table 7.4
Current and Projected Chinese Naval Forces

Chinese Naval Forces	Current Inventory (2015)	Planned Inventory (2025)
Carriers	1	2
Destroyers	27	30
Frigates	45	46
Amphibious warfare ships	52	54
Amphibious warfare craft	305	305

SOURCE: *Global Security*, "Chinese Warships," web page, undated.

be in cases where attack submarines, aircraft carriers, and other surface combatants and aircraft could not be deployed quickly enough to the theater or, if deployed, could not operate close enough to the operating areas of the PLAN. In the Pacific, the constraints on U.S. operations would be greatest in conflicts that occur close to the mainland, where Chinese A2 capabilities are specifically designed to inhibit U.S. power projection. In these circumstances, U.S. forces might be forced to operate from standoff distances or to target the A2 forces, not the ships.

Broadly speaking, the scenarios in which both the location and the timing stress the U.S. capability for maritime interdiction are those where the United States intervenes in a war between an ally and a great power. Examples might include a Chinese invasion of Taiwan or a Russian invasion of the Baltics accompanied by amphibious operations in the Baltic Sea. A conflict between China and Japan over the Senkakus is somewhat similar, but the demands for anti-ship capabilities in a Senkaku scenario would presumably be far less than in a large-scale invasion by a great power and, thus, easier to meet. Only in the case of a war with a great power would U.S. maritime interdiction capabilities be so constrained. In that case, predeploying land-based ASMs could increase or sustain the initial flow of anti-ship capabilities.

In other conflicts situated farther from the Chinese mainland, such as in the South China Sea, the PLAN would be forced to operate outside the protection of land-based air defenses and other A2/AD capabilities. Additionally, not all the PLAN forces discussed in Table 7.4 could project into the South China Sea. This means that U.S. air- and sea-based forces could more easily arrive and operate with fewer constraints, in principle unleashing more of their existing anti-ship capabilities. Thus, in these scenarios, land-based ASMs deployed by the United States would likely not provide significant additional capability.

Thus, the demand for land-based anti-ship capability to conduct longer-range maritime interdiction will be principally limited to scenarios near the Chinese mainland, where Chinese A2/AD capabilities are greatest. Taiwan and Japan could use such systems to protect against attacks. Figures 7.3 and 7.4 depict the locations that fall within 300 kilometers and 500 kilometers, respectively, of the center

Figure 7.3
Potential Basing Locations Within 300 Kilometers of Two Flash Points

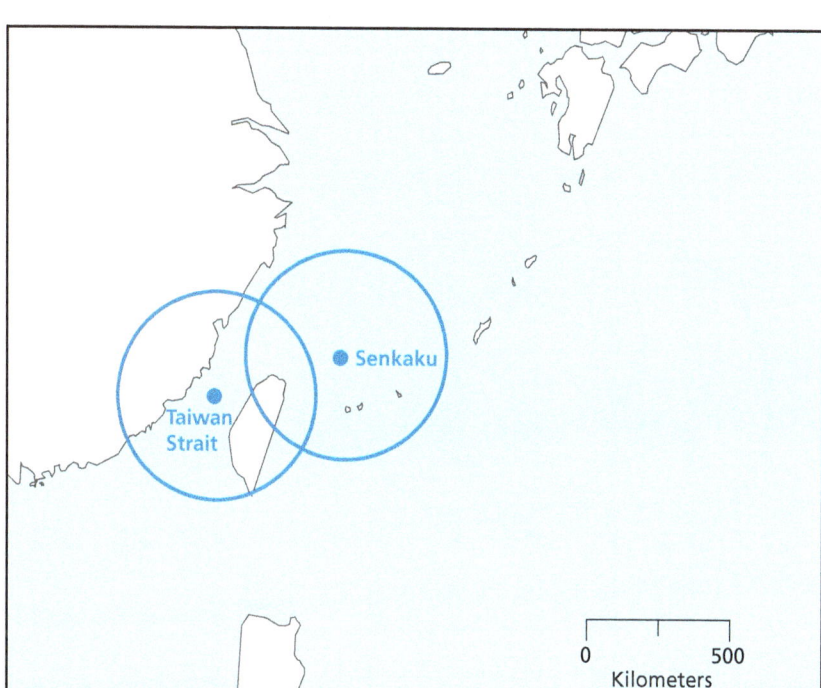

point of the Taiwan Strait and the Senkakus—two potential flash points. The Joint Strike Missile and the ATACMS are two systems in the 300-kilometer–range class,[5] while the LRPF system and the Long-Range Anti-Ship Missile (LRASM) are in the 500-kilometer–range

---

[5] The Joint Strike Missile is a derivative of the Norwegian Naval Strike Missile under development by Raytheon and Kongsberg for the U.S. Navy. The Naval Strike Missile has a range of 200 kilometers; the Joint Strike Missile has a planned range of close to 300 kilometers. Although the Joint Strike Missile is designed specifically for air launch from the F-35, a ground-launched version should be possible if sufficient demand exists. See Raytheon, "Naval Strike Missile: Adaptable, Long-Range Precision," 2016; *Defense Industry Daily*, "Kongsberg's NSM/JSM Anti-Ship and Strike Missile Attempts to Fit in Small F-35 Stealth Bay," November 12, 2015.

Figure 7.4
Potential Basing Locations Within 500 Kilometers of Two Flash Points

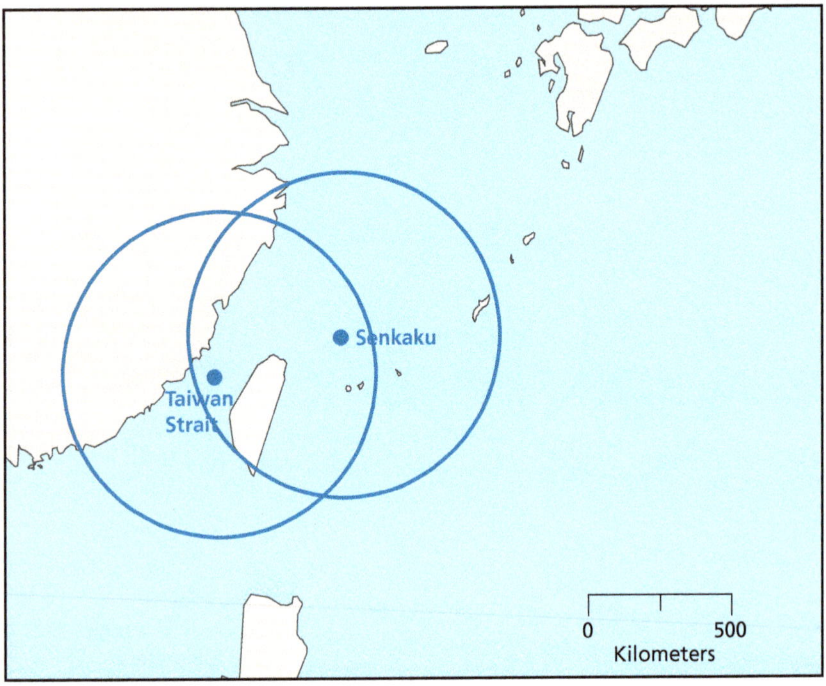

class.[6] Taiwan and Ishigaki stand out as sizable landmasses within a 300-kilometer range of the center point of the Taiwan Strait. Okinawa is farther than 500 kilometers from the Taiwan Strait. The Senkakus are proximal to the Ryukyu Islands, which opens up more options for basing, including Okinawa.

Land-based ASMs in these locations would be vulnerable to multiple modes of attack from Chinese forces. All these locations are within range of Chinese strike assets, including short-range ballistic missiles,

---

[6] If anything, a ground-launched version of the LRASM might have too much range, given that the air-launched version is planned to have a range of 500 nautical miles (926 kilometers). This range would need to be shortened—for example, by adding a heavier warhead and utilizing only a small rocket booster—to abide by INF Treaty restrictions. See *Defense Industry Daily*, "LRASM Missiles: Reaching for a Long-Range Punch," May 17, 2016.

land-attack cruise missiles, and armed UAVs. Thus, survivability will be a crucial parameter for weapon systems and the concept of operation. The more operationally valuable the capabilities are to the United States, the more lucrative targets they make for China. Additionally, other than Taiwan, all the locations are in the Ryukyu Islands, which are Japanese territory.

More flexibility would open up if the United States could develop anti-ship weapons that exceed the range limits of the INF Treaty. Of course, this is not a U.S. Army decision; to withdraw from the treaty, the United States would have to show that events "jeopardized its supreme interests," according to Article XV of the original treaty.[7] Still, a land-based anti-ship weapon with a 1,000-kilometer range could reach the Taiwan Strait or the Senkakus from northern Luzon (see Table 7.1), and the Taiwan Strait can be reached from Okinawa. While Luzon would offer considerably more space for such forces to deploy within, it would also require the Philippine government to commit to becoming involved in a conflict over Taiwan.

As mentioned in Chapter Six, ASMs and SSMs with ranges of 300 kilometers to 500 kilometers could control the Baltic Sea from just one or two sites. See Figure 6.9 for the area that missiles could cover from exemplar sites in Poland and the Baltic Sea islands off the coast of Estonia. ASM forces in the Baltic Sea area, deployed before a Russian attack on NATO, could effectively deny the use of this area to the Russian Baltic fleet.

In summary, U.S. allies would benefit from greater land-based anti-ship capability, which would help in circumstances where employment of U.S. capabilities could be constrained by potent enemy A2 capabilities. The relevant scenarios are exemplified by a Chinese invasion of Taiwan and perhaps by Russian amphibious operations in the Baltics coupled with other attacks on NATO. In such cases, a small U.S. Army unit predeployed within range could complement existing U.S. capabilities, both early in a fight, when flow of forces is limited,

---

[7] U.S. Department of State, Treaty Between the United States of America and the Union of Soviet Socialist Republics on the Elimination of Their Intermediate-Range and Shorter-Range Missiles (INF Treaty), December 8, 1987.

and later. The basing options, however, would be limited in the Pacific, particularly if the United States cannot develop weapons that exceed the range of the INF Treaty and if there is a need for another nation to allow the basing. In all cases, policymakers would have to consider what presumed capability gap a new, potentially expensive Army land-based missile would fill and which targets not currently reachable are so important that they require a new investment by the Army. We develop and cost a concept for a small, predeployed Army unit at the end of this chapter.

Lethal Enforcement of a Blockade

Finally, land-based ASMs could enforce a blockade with lethal fires capable of sinking evading ships. Here, a blockade would be used to prevent access to or use of a maritime area by the military or civilian vessels of a hostile state. Examples could include stopping all ship traffic to or from North Korea if its government collapses to prevent the proliferation or transfer of nuclear weapons and materials to third parties or terror groups. Or a blockade might be imposed on Iranian ports if that nation were to produce nuclear weapons. If Russia invaded the Baltics, a blockade might be imposed on all ships attempting to use the port of Saint Petersburg. A blockade could in theory be imposed on disputed islands, rocks, or reefs in the East or South China Sea. The concept might be that, if such features were seized by China, then Japan or the Philippines might simply declare that any ships attempting to land on, or depart from, the features would be subject to immediate attack without warning. Other authors have assessed the technical feasibility of imposing a general blockade on all of China if it were to attack its neighbors in pursuit of the territorial claims described earlier in this report.[8]

There are open questions about the feasibility and effectiveness of blockades, including the time delay before a blockade would have an effect; the cascading regional and global economic effects from a blockade; and the challenge of discriminating vessels destined for, or originating from, the blockaded area or nation. For this analysis, we assume

---

[8] See Kelly et al., 2013.

that a blockade is a legitimate option, however attractive or feasible that option might actually be. Our focus is on the role land-based ASMs might have in implementing a blockade.

Enforcing a blockade involves multiple steps or tasks. First, vessels need to be detected, tracked, and identified to determine which are potentially attempting to break the blockade. Second, ships that might be attempting to run the blockade must be interdicted. Third, those ships presumably must be boarded, searched, and perhaps seized by forces authorized to use lethal force. Fourth, when search and seizure is deemed insufficient or is not possible, there might be a need to enforce the blockade with disabling fires—for example, by shooting into the bridge or engine room, which can be a technically nontrivial task. The final measure, if the preceding actions prove insufficient or infeasible, would be enforcing the blockade by sinking the ship or ships.

The function of land-based missiles in a blockade is to threaten or attack with the intent to sink a ship attempting to run the blockade. The specific weapon systems are not critical here; in principle, either short- or long-range ASMs could be used for this purpose. Land-based missiles might have some utility in a blockade under a limited set of circumstances:

- *Partners employ anti-ship weapons or grant bases for U.S. forces to do so.* There is no U.S. territory in these scenarios from which to execute a land-based blockade given the INF Treaty limitations. The risk for partners is that using such missiles, or allowing them to be employed from their territories, risks attack by the blockaded nation or retribution after the conflict.
- *The rules of engagement (ROE) allow for the use of lethal force up to and including sinking ships.* More-restrictive ROE, such as those requiring blockaded ships to yield to boarding and search operations, require unambiguous orders to stop ships and to warn of the lethal consequences of not doing so. This typically requires the blockading side to employ naval craft to intercept, board, and search suspected blockade runners. Although land-based missiles could threaten blockade runners with lethal consequences for refusing to stop, using the missiles without first providing

suspected blockade runners with unambiguous warning and an opportunity to submit to a search could be viewed as a disproportionate use of force. Unrestricted shoot-on-sight orders are unlikely, unless war is already under way. It is particularly important to note that the blockading party must be willing to apply force against military, paramilitary (e.g., ships from the coast guard or other law-enforcement agencies), and civilian vessels of any nation entering the blockaded area.

- *Naval forces are unavailable or insufficient to provide lethal enforcement.* Traditionally, naval forces would be the preferred means of executing a blockade, because they offer the graduated enforcement options listed above. While some partners and allies lack the numbers and types of ships needed, the U.S. Navy has the capability and trains to execute all phases of a blockade. Additionally, the U.S. Navy can operate from international waters and, unlike land-based forces, does not need the permission of partner nations. For the U.S. Navy to be disfavored, there must be higher-priority missions for it to execute, which could happen in a broader campaign or a war with a great power. In these cases, the U.S. Air Force could also present formidable anti-ship capabilities by integrating current or future missiles with existing bombers.

These conditions may be most likely to occur when war is already under way between a U.S. ally or partner and some hostile nation. If that ally or partner is operating alone, then ASMs might contribute to a blockade operation being conducted under the conditions listed above.

If the United States has joined with an ally in a war, then land-based missiles might be helpful in cases where U.S. naval and air forces themselves were not available in sufficient numbers at all the times and places needed to conduct this mission. Given the size and power of U.S. naval and air forces, this seems unlikely, unless, as noted, those forces are already consumed by other operations.

This could happen in the case of a U.S. war with a great power. For example, if Russia were to attack the NATO Baltic states, then the U.S. Air Force could be completely consumed with combat operations in Europe and defending other U.S. allies and bases around the periphery of Russia. Similarly, the U.S. Navy might be stressed by the demand to conduct anti-submarine patrols and escort ships moving equipment and supplies to Europe. In such cases of maximum U.S. military efforts, land-based missiles might reduce the need for naval and air forces in some maritime-control operations.

However, blockade missions in times of war merit some additional considerations. First, the blockade would be nested in a broader campaign likely to include attacks on the adversary's high-value naval vessels. The U.S. Navy and Air Force might already have sunk or damaged many of the enemy naval vessels sailing in the region.[9] The implication is that a small number of adversary naval vessels, and perhaps none, remain as potential targets for the land-based ASMs.

Second, the operational concepts and ROE would have to address civilian vessels. This could include warning by lightly armed friendly vessels carrying law enforcement or naval infantry for boarding, search, and—as needed—seizure and impounding. The ROE, however, would need to allow land-based missiles to employ lethal force if a civilian ship attempted to evade or resist. Some naval forces, albeit light ones, would still be required in this concept unless the United States and its allies were willing to extend shoot-on-sight rules to halt civilian traffic. This seems unlikely except in the direst of circumstances.

Potential Army Roles in Long-Range Ground Strike Operations

In this section, we consider land-force contributions to long-range ground-strike missions across ocean areas. First, we discuss existing and planned SSMs, such as the ATACMS and the LRPF system, and we examine the potential roles for SSMs with ranges that exceed

---

[9] However, the United States might limit strikes on adversary naval and air forces that remain in their homeports. For example, the United States might choose not to strike Russian naval vessels in Saint Petersburg for fear of escalating a war in Europe. (One could imagine Russia launching cruise missile attacks against U.S. Navy vessels in Norfolk or San Diego in retaliation.) It may be more palatable to sink any vessels that emerge from homeports if long-range ASMs can be placed in the appropriate positions to blockade them.

500 kilometers. Then we examine concepts for inserting weapon systems from the existing Army inventory into denied environments. In each case, we assess the demand for land-based strike in the context of a high-end conflict, such as an invasion of Taiwan or a large-scale conflict emanating from the South China Sea; otherwise, the Navy and Air Force would have ample assets available without the need for ground-based systems.

Long-Range Surface-to-Surface Strike

Here, we assess the particular characteristics of *land-based* SSMs and the challenge they present to an opponent searching for the launchers in the field. There are plausible demands for strike missions as part of a broader campaign that may involve disrupting adversary A2/AD capabilities. Elements of Chinese or Russian forces with A2/AD capabilities are obvious potential targets early in a conflict.[10] Potential targets would include integrated air defense systems, air bases, surveillance systems, and enemy long-range strike systems. Of course, the joint force has significant long-range strike capabilities, with bomber fleets, land- and sea-based strike fighter aircraft, and air- and sea-launched cruise missiles.

The differentiating feature of an SSM is the delivery mechanism. A ballistic missile arrives quickly, within minutes; aircraft and cruise missiles, while capable of delivering a great deal more payload, arrive many minutes or even hours later. Indeed, ballistic missiles have been considered for long-range, conventional strike at intercontinental distances, beyond the range limit of the INF Treaty; their numbers, however, are limited by other treaties. The appeal of U.S.-owned long-range SSMs has even been recognized in the context of a conflict with China, although they would be launched from naval platforms, which avoid the INF Treaty constraints.[11]

Past and potential adversaries have long recognized the potential benefits of long-range SSMs. Hezbollah and Hamas have employed

---

[10] Jan van Tol, Mark Gunzinger, Andrew F. Krepinevich, and Jim Thomas, *AirSea Battle: A Point-of-Departure Operational Concept*, Washington, D.C.: Center for Strategic and Budgetary Assessments, 2010.

[11] Van Tol et al., 2010, p. 83.

ballistic missiles against Israel, and Iraq has employed them in two wars against the United States. In all these cases, the ballistic missiles had very limited military success but a good deal of psychological effect. Perhaps most notably, fixed and mobile launchers proved to be very difficult to find and kill before they launched their missiles. In the cases of Operations Desert Storm and Iraqi Freedom, the United States conducted a massive effort to find and kill launchers on the ground and missiles in flight.[12] North Korea could pose an even greater threat with its long-range artillery rockets and even longer-range ballistic missiles. China and Russia would pose still greater threats to U.S. and allied operations.

Today, the U.S. Army operates the ATACMS with a range of 300 kilometers and is developing the LRPF missile with an INF Treaty–compliant range of 499 kilometers.[13] The new missile would most likely be issued to existing ATACMS units. Both of these missiles could be useful to disrupt adversary operations.

Within this context, it is unlikely that a new very long-range SSM would be procured simply to augment the existing joint strike capacity.[14] The expense of developing and fielding a new long-range SSM would be significant, both in absolute terms and in comparison to the cost of improving existing joint or Army force structure. New weapon

---

[12] For an excellent description of U.S. air operations to attack Iraqi air defenses and ballistic missiles—an early form of an A2/AD system—see James A. Winnefeld, Preston Niblack, and Dana J. Johnson, *A League of Airmen: U.S. Air Power in the Gulf War*, Santa Monica, Calif.: RAND Corporation, MR-343-AF, 1994. Several documents give an excellent description of the operations against Iraqi air defenses and ballistic missiles in Operation Iraq Freedom and an account of U.S. air and missile defense operations. See 32nd Army Air and Missile Defense Command, *Operation Iraqi Freedom Theater Air and Missile Defense History*, Fort Bliss, Tex., September 2003; Michael T. Moseley, *Operation IRAQI FREEDOM— by the Numbers: Assessment and Analysis Division*, Shaw Air Force Base, S.C.: U.S. Central Command Air Forces, April 30, 2003; Gregory Fontenot, E. J. Degen, and Dave Tohn, *On Point—The United States Army in Operation Iraqi Freedom*, Fort Leavenworth, Kan.: Combat Studies Institute Press, 2004.

[13] See Sydney J. Freedberg, Jr., "New Army Long-Range Missile Might Kill Ships, Too: LRPF," *Breaking Defense*, October 13, 2016.

[14] An exception might be in those cases in which an improvement to an existing system could provide significant advantages. For example, missiles that improve on the range of the ATACMS system, such as the LRPF system, may be very useful for conflicts in Europe.

systems and, for a land-based system, new Army force structure would need to be developed, procured, trained, and fielded. This means that the new system would have to provide a special capability and would most likely be fielded as a high-value, low-density asset and not used to strike "just any old" target.

Nonetheless, a long-range SSM could have three competitive advantages through its *capability*. The first competitive advantage would be *responsiveness*. Figure 7.5 shows the results of basic ballistic calculations that provide responsiveness at different ranges for different strike platforms and munitions.[15] These calculations show that the time of flight of an SSM that is roughly 1,500 kilometers from its target would be on the order of ten to 20 minutes. Responsiveness of subsonic strike platforms and cruise missiles would be measured in hours. Of course, responsiveness of subsonic missiles could be augmented if the platforms that carry them (e.g., bombers and submarines) are on station near the target, but such augmented responsiveness requires a

Figure 7.5
Depth of Attack and Responsiveness of Strike Assets

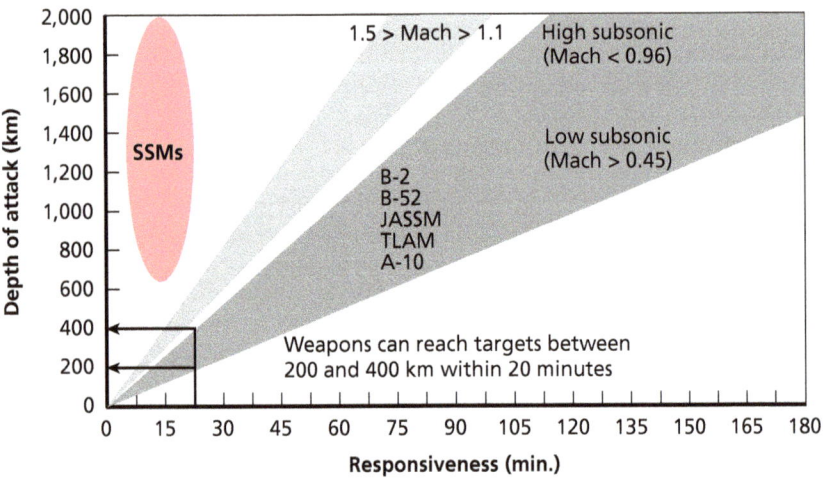

---

[15] C2 delays and launch delays are not included in these calculations.

large number of orbits to cover many targets with a larger number of platforms and also requires operating those platforms deep within the adversary's A2 envelope to compensate for the slow flight times. Additionally, these orbits would often require additional supporting aircraft to defend or enable the strike platforms by providing surveillance.

As one illustration, Figure 7.6 shows that roughly nine air-launched strike platforms—loitering within or near the range of air defenses—would be needed to guarantee 20-minute responsiveness across the area covered by long-range SSMs, if the missiles were based in the northern Philippines.

The advantage of responsiveness could be used very early in a conflict if the capabilities were prepositioned, ready to launch on cue, and akin to the concept of Prompt Global Strike that DoD has examined.[16] Such a concept, assuming it would be operationally effective, is well suited to submarine SSMs, because submarines are covert and difficult to suppress.

Figure 7.6
Response Time and Coverage of Airborne Versus Land-Based Strike Assets

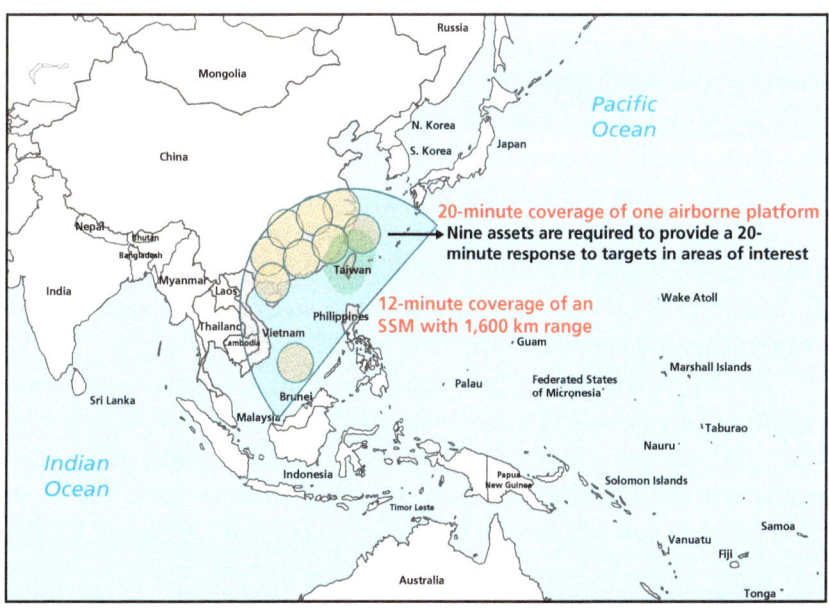

It could be difficult, however, to have land-based prompt strike prepositioned in theater. This would require, well ahead of a conflict, an ally or partner (such as the Philippines) to agree to host U.S. forces able to launch weapons from its territory. Few nations in the western Pacific seek conflict or even tension with China. Additionally, as has been shown with the positioning of Terminal High Altitude Area Defense (THAAD) in South Korea, China would likely consider such deployments provocative in and of themselves, potentially leading to a crisis instead of deterring one. Prepositioning launchers and support equipment while deploying missiles only in a crisis or conflict might be able to mitigate these destabilizing effects. In this case, a land-based prompt *regional* strike capability would need to yield operational effects within a broader campaign—not just in an initial quick strike, as sometimes envisioned in the Prompt Global Strike concept.

A second competitive advantage of a long-range SSM would be an *ability to penetrate active defenses*. The U.S. experience developing both theater and strategic ballistic missile defenses demonstrates the large technical hurdles that must be addressed; as others have noted, the United States' active defenses are much more costly than the missiles they are defending against.[17] The high velocities of ballistic missiles challenge the engagement timelines of a defense system—that is, detection, discrimination, and tracking must be accurate and quick for the interceptor launch to be effective—while their long ranges demand long-range radars for detection and tracking and long-range interceptors for increasing the kill probability. Adversaries would be forced to address those same challenges if they developed active defenses to counter a U.S. ballistic missile capability and to deploy defenses across all the targets at risk. Again, this is an advantage of SSMs, regardless of whether they are launched from land or sea.

Finally, a land-based SSM offers a third advantage—*enduring availability*. In principle, a few such missiles can remain ready in the field to launch throughout a conflict, with assured, comparatively low-bandwidth communication. Such dispersed missiles are difficult to locate, as the United States discovered in Iraq during Operation Desert

---

[17] Van Tol et al., 2010, p. 36.

Storm and again during Operation Iraqi Freedom.[18] In contrast, maintaining naval surface forces or, more likely, submarine forces on station to provide a few missiles is more difficult, because the opportunity cost is larger and because the risk is greater for surface forces.

Translating these technical advantages of responsiveness, ability to penetrate active defenses, and enduring availability into operational and campaign-level effects requires an operational concept that exploits them. One such developed operational concept is a long-range sniper that would strike fleeting and potentially heavily defended targets of opportunity. We discuss the operational concept below.

Long-Range Sniper Concept

The long-range sniper concept would have mobile SSM launchers prepositioned far outside the range of A2/AD capabilities prior to conflict. Personnel for the unit, and perhaps missiles, would flow into theater after the start of a crisis. During a campaign, a few SSMs would be kept on alert—perhaps three to six. The alert SSMs would be launched against fleeting targets when intelligence detected a target, such as a mobile command center or radar, or against an activity that the missile could usefully disrupt, such as a large air strike being assembled involving multiple airfields.

Just as with a tactical sniper, the opportunities for such attacks would be partly unpredictable; for example, not all enemy operations massing aircraft for an attack would be detected in a timely way. But, also as with a tactical sniper, the effect on the opponent would be real and possibly large. Beyond having to provide defenses against such ballistic missiles, an adversary might have to change the way it operates command centers or organizes attacks, because of the possibility of such ballistic missile attacks. Additionally, the temporal precision of the strike, when it is possible, might reduce the need for other strike missiles and platforms. To continue the previous example, if the United States has confidence in detecting the assembly of a strike formation, then striking an air base precisely when strike packages are

---

[18] Timothy M. Bonds, Eric V. Larson, Derek Eaton, and Richard E. Darilek, *Strategy-Policy Mismatch: How the U.S. Army Can Help Close Gaps in Countering Weapons of Mass Destruction*, Santa Monica, Calif.: RAND Corporation, RR-541-RC, 2014.

being formed may be more efficient than sustaining suppression over a long period to ensure that strikes cannot be generated. The SSM attacks might even create disruptive effects that cascade disproportionately through the campaign. This is distinct from other strikes, which provide destructive effects from the sheer volume of fire. Such effects are likely real, but their operational value is difficult to estimate.

Naturally, such a concept requires a place to base and stage the missiles and launchers. Different sites have different implications. Hawaii and Guam are the two potential locations for basing a theater strike capability from U.S. territory, and both are attractively distant from Chinese A2 threats. But they are more than 8,000 kilometers (Hawaii) and 2,500 kilometers (Guam) away from potential target locations (shown in Table 7.1). SSMs in either location would be less responsive and, perhaps more important, significantly more costly. The northern Philippines stands out as a uniquely attractive basing option from a purely operational perspective. Luzon is a ten- to 20-minute flight from most relevant target locations yet still sufficiently distant from A2 threats. The northern Philippines also offers a large operating area and terrain for concealment, enhancing the survivability discussed above. Kyushu presents another option with similar advantages to the northern Philippines and even better infrastructure, but it would be relatively distant from the South China Sea.

Importantly, both the northern Philippines and Kyushu have the inherent disadvantage of requiring partner approval for access and prepositioning. And neither the Philippines nor Japan has an obvious reason to become involved in countering an invasion of Taiwan. Nonetheless, any Army system would likely need to be located in one or the other nation.

The operational effects described here are hypotheses. They have not been validated by analysis and are thus insufficient to justify investments or withdrawal from treaty obligations. A more detailed campaign analysis would be needed to examine this hypothesis, and perhaps others, in terms of the potential operational benefits of long-range, land-based SSMs.

The cost of developing and fielding such a capability would be significant. A rough estimate suggests such a capability could cost more than $4.5 billion to develop and between $1.0 billion and $1.5 billion to procure equipment to field a Pershing-II–type battalion. Similar to the Pershing II, this new capability could take as long as a decade to field. The relatively high costs would also create pressure to make the weapon multi-use—for example, field-swappable payloads that would allow tailoring the warhead to different target types.

Insert Existing Capabilities into Denied Environments

Another option to increase the contribution of surface-to-surface fires is to develop concepts to rapidly insert fire units equipped with weapon systems in the current Army inventory, such as ATACMS and the Guided Multiple Launch Rocket System (GMLRS), within range of relevant targets. In principle, this option would change the calculus above by making the current weapon relevant even if it cannot be pre-positioned. As a practical matter and as we discuss below, the limited range of ATACMS and GMLRS necessitates inserting these units into environments where a capable adversary intends to deny access. This option would require forces ready to deploy in a crisis, as well as concepts that allow for Army units to be integrated into Air Force or Navy packages for protection during insertion. The option might also require developing special payloads that increase the effects of the small force that could be deployed in this way.

As discussed, Army offensive fires are inherently limited in range, currently reaching less than 300 kilometers, even when using ATACMS—the longest-range system in the U.S. inventory. Historically, Army field artillery has searched for ways to compensate for the always-present limitation in range so the artillery could attack deeper targets. One idea for allowing deeper attacks is an artillery raid, described in 1999 as the following: "Conduct an air insertion of a firing battery (+) in a cross-FLOT [forward line of own troops] operation to attack designated enemy HPTs [high-priority targets] with lethal indirect fires. Be prepared to extract the firing battery (+)."[19] A more recent

---

[19] Federation of American Scientists, Annex G, "APP 7—Artillery Raid, Tactics, Techniques, and Procedures for Air Assault Operations," March 17, 1999.

concept has been developed and exercised by elements of I Corps. That concept is discussed below.

I Corps HI-RAIN Concept

The basic concept investigated by I Corps is called HI-RAIN. HI-RAIN uses HIMARS in combination with transport aircraft to perform an artillery raid. The name HI-RAIN comes from **HI**MARS **Ra**pid **In**filtration. It is described as a forcible-entry capability that combines either a C-130 or a C-17 with one or two HIMARS launchers. HI-RAIN proponents at I Corps argue that the concept can serve as an operational or even strategic strike package in support of a joint task force commander.[20] HI-RAIN is an extension of previous artillery raid concepts, dating back to at least the late 1990s. The general idea for all artillery raids is the same as for HI-RAIN: "Raid missions support mission objectives by sending firing elements forward to engage enemy HPTs [high-priority targets] that are currently beyond the maximum range of the HIMARS weapon system."[21] Schematically, the concept is illustrated in Figure 7.7.

In this concept, the artillery system, in this case HIMARS, is moved beyond the forward line of own troops into territory that is, by definition, not controlled by friendly forces. In HI-RAIN, this is typically accomplished by one or two transport aircraft carrying HIMARS and landing at some available landing field. The artillery then disembarks, sets up, and fires at targets—the red diamonds in Figure 7.7— that were previously out of range of the artillery. The goal is to achieve some element of surprise. In more detail, the concept relies on some

---

[20] Mark Miranda, "Artillery, Air Crews Execute HI-RAIN Joint Exercise," U.S. Army, February 6, 2014. A closely related concept, Hot Panel, is used by other units, including the 18th Field Artillery Brigade at Fort Bragg and the 157th Field Artillery Regiment, which is in the Colorado National Guard. The difference is that, unlike HI-RAIN, Hot Panel does not rely on the Joint Precision Airdrop System GPS Retransmission System within the transport aircraft. Use of that system in HI-RAIN allows the HIMARS to be powered down until it is near the target area but then still acquire its GPS position before landing, allowing a longer flight time while still not requiring GPS acquisition after exiting the transport aircraft. This speeds the execution of the artillery raid.

[21] 62nd Airlift Wing Public Affairs, "Airmen, Soldiers Train to Rapidly Infiltrate Enemy Territory," Air Mobility Command, June 28, 2016.

**Figure 7.7
HIMARS Raid Concept**

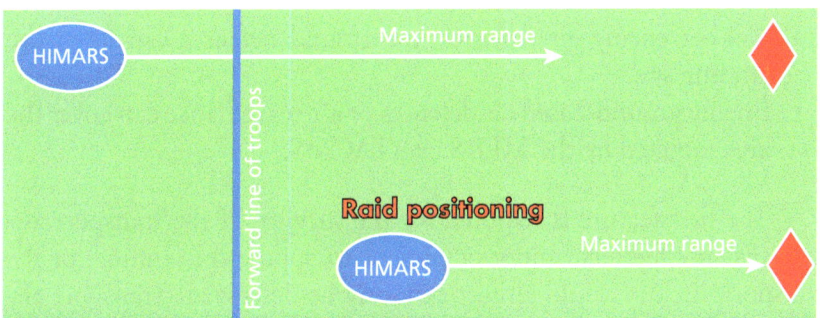

SOURCE: Federation of American Scientists, 1999, slide 4.
RAND RR1820-7.7

smaller systems developed for other purposes—for example, the GPS retransmission system used for the Joint Precision Airdrop System. GPS retransmission systems within the cargo and crew areas of a transport aircraft allow GPS receivers inside the HIMARS to know their location before landing.[22]

HI-RAIN itself depends on the ability of transport aircraft, such as C-17s, to penetrate the air space of interest, land, and then reload and return with the HIMARS systems and crews. In theory, an alternative might be to resupply the HIMARS if it could continue to operate in the area. The collateral implications of the concept have important implications for balancing the risks of this concept against the risk inherent in other potential operational concepts. Because the raid envisions transport aircraft landing within about 70 kilometers of a target (for GMLRS) or about 280 kilometers (for ATACMS), several critical features of the operational environment are implied, the most important of which appear to be the following:

---

[22] The concept has been practiced and embedded in exercises, exploiting the convenient colocation of the 62nd and 446th Air Wings with the 17th Field Artillery Brigade at Joint Base Lewis McChord (JBLM). Integration with special operations forces has also been practiced with the 2/75th Ranger Battalion, also at JBLM.

- Allied or U.S. forces must have assured air superiority over the route of flight and the landing site for the transport aircraft, whether it is a C-17 or C-130.
- Neither enemy air defenses nor ground defenses can be at the landing site.
- Enemy ground-based air defenses or air surveillance can cover the area targeted by the MLRS or ATACMS.

The first feature is required for the survival of the transport aircraft. Enemy fighters cannot be present to oppose the landing, or the transport aircraft would almost certainly be destroyed. Transport aircraft cannot avoid detection from a modern fighter, even at night, nor do they have effective defenses. This condition is readily attainable in a generally permissive environment—for example, when the United States is opposing an insurgent force with no air force. Against an opposing nation-state though, this would usually require a fighter escort for the air transports. The escort is necessary because even the range of ATACMS, about 300 kilometers, is small compared with most tactical fighter ranges, so the HI-RAIN raid will not typically simply out-range an opponent's air force. Notably, the personnel at risk in a HI-RAIN raid are roughly 15 if two HIMARS are used, which is significantly larger than the two to five at risk on a bomber sortie.[23] This magnifies the significance of any risk to the transports.

The second feature is again necessary for the survival of the raiding force. Ground-based air defenses can easily damage a transport aircraft, and even very short-range ones at the landing site would threaten a transport. Similarly, while a security detachment is part of the HI-RAIN concept, it is small, on the order of four soldiers, because of the need to use few transport aircraft to quickly execute the raid. Since transport aircraft on the ground are vulnerable to many direct fire weapons, the landing area cannot have significant enemy forces or be within range of enemy artillery that could attack the aircraft on the ground. The last condition could be dropped in cases where the entire HI-RAIN package and all supplies can be air-dropped as opposed to air-landed.

---

[23] Information in this paragraph and the next was conveyed to RAND by the 17th Field Artillery Brigade.

As to the third feature, the United States has many heavy bombers capable of dropping conventional weapons—B-52s, B-1Bs, and B-2s. Without some threat to a bomber, or at least warning that it was arriving in the area, a direct attack from a bomber would be as timely and as accurate as an MLRS or ATACMS barrage and could even easily scale to larger aggregate attacks. Therefore, air defenses above the enemy target must be present to make HI-RAIN clearly preferable to the obvious alternative, a bomber delivering a Joint Direct Attack Munition or other ordnance. Even the presence of some air defenses might not tip the choice to HI-RAIN versus a bomber, because bombers do have some countermeasures and risk fewer personnel. Additionally, bombers or ships can launch cruise missiles to attack many targets with little risk to personnel, albeit with a much longer delay from launch to impact on the target. In a fully permissive environment with no air defenses at all, the bomber would face essentially no risks, unlike HI-RAIN, which must find a secure landing area. This makes bombers more attractive for most, but perhaps not all, COIN or counterterrorism strikes.

Another concept derived from HI-RAIN could be applicable to certain conflicts with China. This concept would involve inserting a small U.S. force equipped with HIMARS to support an ally involved in a large-scale ground conflict with China. Like the prior concept, this differs importantly from HI-RAIN in having the deployed HIMARS remain deployed, and resupplied, probably intermittently by airlift. And the concept differs in that the ally would be responsible for ground security near the U.S. HIMARS. It also differs in that the actual operational area for the HIMARS would shift tactically and operationally through the campaign and not remain tied to a single landing site.

Those employing this concept must assume that a capable adversary would often be able to contest air superiority over the landing areas. Performing the insertion in conjunction with some other large U.S. air operation would then seem to be necessary to drive away or at least distract the opposing aircraft from the vulnerable transport aircraft carrying the HIMARS and a C2 node. As with HI-RAIN, there would also be a great priority in having the HIMARS quickly exit the transport, before enemy aircraft might react. Additionally, resupply would be limited in number and frequency. Given the possibility of coordinating resupply with other U.S. air operations, such reinsertion and resupply seem to be plausible.[24]

To be effective in such support to an ally, the U.S. forces would need to provide some important capability not otherwise available to the ally and would need to be effective using only a small number of rounds. Any high volume of fires, MLRS or otherwise, would remain the responsibility of the ally and would need to be prepared far in advance of the conflict. Conceptually, there could be special payloads—for example, employing some electronic warfare capability—especially useful at critical moments in a campaign. Typically, if such systems were developed, they would be closely controlled and, as such, would have to be provided by the United States.

Overall, this concept would be attractive only if several factors are present. First, the ally in question must have a reasonably capable military; otherwise, this sort of capability cannot make enough of a difference. Second, at least occasional air transport must be possible near the area of conflict, perhaps involving a fighter escort or diversionary attacks while the transports land. Third, the United States must actually develop such special payloads that would have an unusual effectiveness and also not be otherwise available to an ally. Finally, agreement, training, and exercises with the ally must be made well in advance of any conflict; otherwise, the coordination for such insertion, resupply, and the effective use of the special fires would not be possible.

When these factors are present, though, this concept, derived from HI-RAIN, appears to offer a military capability not otherwise available. U.S. or friendly aircraft cannot simply loiter in such conditions, which would be needed for being responsive to a tactical ground battle.

To pursue this possibility, the Army would need to do more than practice HI-RAIN. The special munitions must be successfully developed and tested, HI-RAIN–type insertions should be practiced in conjunction with larger air operations, and the necessary arrangements must be made with the prospective ally. Fortunately, none of

---

[24] U.S. Air Force, "MC-130E/H Combat Talon I/II," fact sheet, March 28, 2003.

these preparations is unusually expensive, which allows the Army to easily prepare for this option if desired.

Potential U.S. Army Roles in Defending Against Low-Altitude Air and Cruise Missile Attacks

Among the most sobering of Chinese and Russian military capabilities is the countries' extensive preparation for A2/AD strategies.[25] Although much public attention has been focused on China's growing ballistic missile force, this is only part of its capability to strike targets at significant ranges from China. The future Chinese military will have an extensive set of combat aircraft and cruise missiles to complement its ballistic missile force. Similarly, Russia's fortification of the Kaliningrad enclave with advanced surface-to-air, anti-ship, and long-range rocket artillery is well documented.

Chinese military doctrine envisions large-scale, structured attacks combining these forces. A structured attack allows the ballistic and cruise missiles to suppress many air defenses, so that Chinese aircraft can then efficiently deliver large amounts of ordnance to the targets.[26] The U.S. military has noticed these capabilities and the threat they present to the Navy and Air Force. In both cases, the threats are to the bases (carriers or air bases) needed to prosecute the conflict. And, not surprisingly, both services are developing plans to protect their bases.[27]

---

[25] Andrew Krepinevich, Barry Watts, and Robert Work, *Meeting the Anti-Access and Area-Denial Challenge*, Washington, D.C.: Center for Strategic and Budgetary Assessments, 2003; Roger Cliff, Mark Burles, Michael Chase, Derek Eaton, and Kevin L. Pollpeter, *Entering the Dragon's Lair: Chinese Antiaccess Strategies and Their Implications for the United States*, Santa Monica, Calif.: RAND Corporation, MG-524-AF, 2007.

[26] Cliff, Burles, et al., 2007, p. 63; Roger Cliff, John F. Fei, Jeff Hagen, Elizabeth Hague, Eric Heginbotham, and John Stillion, *Shaking the Heavens and Splitting the Earth: Chinese Air Force Employment Concepts in the 21st Century*, Santa Monica, Calif.: MG-915-AF, 2011, pp. 106–107, 184–186.

[27] Marcus Weisgerber, "Pentagon Debates Policy to Strengthen, Disperse Bases," *Defense News*, April 13, 2014; Ronald O'Rourke, *China Naval Modernization: Implications for U.S. Navy Capabilities—Background and Issues for Congress*, Washington, D.C.: Congressional Research Service, September 8, 2014.

For the Air Force, these plans revolve around making any air bases that the United States might use in the conflict more robust against these large, structured attacks. In practice, these plans rely on combinations of greater dispersal, both across different bases and within each base; hardening and redundancy of the bases and their infrastructure; repair capability on the bases; and defenses for the bases.[28] Similarly, ground forces employing long-range fires may be vulnerable to aircraft that can loiter overhead to find targets.

The Army can make important contributions to many of these efforts to increase robustness. Indeed, defending against potential attacks is already a focus of planning in U.S. Army Pacific and the 94th Army Air and Missile Defense Command. Here, we suggest a particular approach, one enabled by important air and missile defense systems that the Army has invested in and that will be deployed in the next five years. These include the Integrated Air and Missile Defense (IAMD) Battle Command System (IBCS) and the first block of the IFPC-2–Intercept (IFPC-2-I), which itself relies critically on IBCS.[29] Together, these systems offer a new air and cruise missile defense capability that we believe is extremely useful in these potential conflicts. If the systems are employed as we describe, we believe that they would greatly complicate any efforts to attack air bases or to suppress artillery fires. When used with other operational responses, these defenses could even change an adversary's fundamental calculus on the outcome of a conflict.

Operationally, the key system here is IFPC-2-I. As we will show, IFPC-2-I offers a number of advantages:

---

[28] Brent Thomas, Mahyar Amouzegar, Rachel Costello, Robert A. Guffey, Andrew Karode, Christopher Lynch, Kristin Lynch, Ken Munson, Chad J. R. Ohlandt, Daniel M. Romano, Ricardo Sanchez, Robert S. Tripp, and Joseph Vesely, *Project AIR FORCE Modeling Capabilities for Support of Combat Operations in Denied Environments*, Santa Monica, Calif.: RAND Corporation, RR-427-AF, 2015; Jeff Hagen, Forrest E. Morgan, Jacob Heim, and Matthew Carroll, *The Foundations of Operational Resilience: Assessing the Ability to Operate in an Anti-Access/Area Denial (A2/AD) Environment: The Analytical Framework, Lexicon, and Operational Resilience Analysis Model (ORAM)*, Santa Monica, Calif.: RR-1265-AF, 2016.

[29] U.S. Army, "Indirect Fire Protection Capability, Increment 2–Intercept (IFPC Inc 2-I)," web page, undated.

- inexpensive proliferation, particularly of the radars, including through Foreign Military Sales
- very deep magazines for countering large air or cruise missile raids
- a favorable cost-exchange ratio between the planned interceptor and aircraft or cruise missiles
- easy deployment
- high political acceptability, when compared with large and more-visible systems, such as Patriot.

Army air and missile defenses can play an important role in protecting U.S. air and long-range fires in conflicts with adversaries that have a capable air force. But, as we have shown, that cruise missile defense must be different from the air and missile defense concepts that have worked well in U.S. Central Command. The scale and sophistication of the U.S. opponent requires different operating concepts. In the western Pacific, air and missile defense can only work in concert with a much greater dispersal and hardening of air bases than in the past.

The defense systems themselves would need to be operated differently as well. Patriot is likely to need to "hide" without radiating, thus preserving it for "popping up" and countering aircraft attacks. IFPC-2-I—using multiple, dispersed, redundant Sentinel radars, all connected by IBCS—is not defending maneuver forces in this concept. Rather, for this concept, it is deployed to provide point defense against low-altitude air or cruise missile attacks. Ideally, the presence of allied and partner Sentinel radars keeps the enemy forces from even knowing where the IFPC-2-I defenses are.

All this would require changes in the IFPC-2-I force implied by this concept. It would be a larger force than the Army's existing program, particularly in the total interceptor inventory, and would need to be either already deployed in the theater or rapidly deployed to the theater in the event of a crisis. This may have implications for the active-reserve mix for the force. The doctrine and training would also need to change to reflect this role in defending fixed sites as opposed to primarily defending maneuver forces. Finally, the IFPC-2-I forces would need to be separate from, or at least easily separable from, the brigade combat teams.

The opportunity would be an Army air and missile defense force that could truly change the balance *without* a great increase in the much more expensive Patriot-class defenses. In a very real sense, this would allow IFPC-2-I to substitute for additional Patriots. And if, as we estimate, this concept would indeed change the Chinese calculus on the overall outcome of the conflict, it might help avoid these conflicts in the first place.

Meeting the Army air defense forces employing IFPC-2 requires several characteristics:

- Each platoon must be able to operate independently, with no critical dependence on higher echelons for mission command or specialized logistical support, until later deployments of supporting units.
- The force design for each platoon must have enough redundancy to increase survivability and a magazine depth large enough to mount an operationally significant defense against a sustained attack directed at key air bases or logistics centers.
- The force structure must allow many platoons to be strategically deployed quickly in a crisis and to be easily redeployed within the Pacific.

Results of Ground-Based A2/AD Mission Assessment

Table 7.5 summarizes our findings and conclusions about ground-based missile missions that entail **controlling or denying hostile maritime operations:** *anti-ship operations* (tactical coastal defense, theater maritime interdiction, and lethal blockade enforcement); **denying or defeating hostile ground operations:** *surface-to-surface operations* (long-range strike and fires in denied environments); and **defending against low-altitude air and cruise missile attacks:** *short-range air and cruise missile defense operations*. The first two anti-ship missions are broken out by adversary and ally, while the third anti-ship mission and the other two operations apply across the allies and adversaries.

Table 7.5
Ground-Based A2/AD Missions—Findings and Conclusions

Mission	Potential Adversary: China Allies: Japan and the Philippines	Potential Adversary: China Partner: Taiwan	Potential Adversary: Russia Ally: NATO	Potential Adversary: Iran Partners: Persian Gulf Nations
		Anti-Ship Operations		
Tactical coastal defense	• Demand high • Potential U.S. policy constraints • Natural mission for Japan and the Philippines	• Demand very high • Policy prohibits peacetime U.S. presence • Natural mission for Taiwan	• Demand very high • Red A2/AD limits NATO sea and air forces • Natural NATO and partner mission	• Demand moderate • U.S. joint anti-ship capability high • Mission for Persian Gulf partners
Theater maritime interdiction	• Demand high • Joint anti-ship capability high • Natural mission for Japan and the Philippines • United States could assist and reinforce	• Demand very high • Joint anti-ship capability high • Natural mission for Taiwan • United States might assist • ASMs could arrive faster	• Demand very high • Red A2/AD limits NATO sea and air forces • Natural NATO and partner mission • United States could assist and reinforce	• Demand moderate • U.S. joint anti-ship capability high • Natural mission for Persian Gulf partners • United States could assist and reinforce
Lethal blockade enforcement	• Demand narrow and rare; blockades are enforced by naval forces in most scenarios. Lethal enforcement by land-based ASMs requires multiple unusual assumptions about the conflict and the availability of naval forces.			
		Surface-to-Surface Long-Range Strike Operations		
<500 km strike	• Existing and improved systems useful for suppressing or destroying enemy air defenses, conducting counterbattery, and striking airfields and other high-value targets			
>500 km strike	• Very long-range sniper concept could offer responsiveness and ability to penetrate enemy defenses, but analysis is needed to examine hypothesized effects – Demand plausible but not established • Operational benefit would be required to justify withdrawal from INF Treaty			
Strike in denied environments	• Demand plausible but not established • A2 environment would inherently limit employment			

Table 7.5—Continued

Mission	Potential Adversary: China		Potential Adversary: Russia	Potential Adversary: Iran
	Allies: Japan and the Philippines	Partner: Taiwan	Ally: NATO	Partners: Persian Gulf Nations
Short-Range Air and Cruise Missile Defense Operations				
Point defenses	• Demand very high because of emerging threats and distributed basing • Joint capability and capacity gaps • Army's emerging IFPC-2 could address gap if force designs and force structure accommodate demands for responsiveness and survivability • Munition supply represents significant expense			

Estimates of the Costs to Field Land-Based Missile Systems

In this section, we assess investment options for the U.S. Army in anti-ship batteries, short-range air-defense batteries, and SSM batteries. We provide some rough cost estimates for each of these options.

We considered five different missile investments, as shown below. The first three would add an anti-ship capability to Army fires battalions. The fourth would increase the Army's short-range air defense capabilities, and the fifth would greatly extend the range of Army surface-to-surface fires.

1. Procure military-off-the-shelf (MOTS) foreign system with 200-kilometer range.
2. Adapt ATACMS with 300-kilometer range to add anti-ship capabilities.
3. Adapt a ground-based LRASM with 500-plus–kilometer range to operate from a ground-based launcher.
4. Procure IFPC-2-I as an air defense missile using AIM-9X missiles.
5. Develop an SSM with Pershing II capability—i.e., 1,500-kilometer range.

We next discuss the cost implications of the alternative missile procurement options, as summarized in Table 7.6.

Table 7.6
Estimated Costs for Initial Deployed Battalion

System	Development Cost ($)	Deployed Missiles per Fires Battalion (number)	Initial Procurement Cost ($)	Incremental O&S (deployments and new force structure) ($)
1. Procure MOTS foreign design— e.g., RBS-15 or Naval Strike Missile	—	160	340M	15M per year for incremental deployed O&S costs
– Each of two batteries		80 (160 total)		
2. Adapt ATACMS	400M	160	140M	15M per year for incremental deployed O&S costs
– Each of two batteries		80 (160 total)		
3. Adapt LRASM	200M to 300M	160	1,504M	15M per year for incremental deployed O&S costs
– Each of two batteries		80 (160 total)		
4. Field IFPC-2-I to improve short-range air defenses	607M[a]	1,350	950M	15M per year for incremental deployed O&S costs
– Each of three batteries		450 (1,350 total)		
5. Develop an SSM Pershing II–like ballistic missile	4,500M	40	160M (launcher) 1,000M to 1,400M (missiles)	15M per year for incremental deployed O&S costs; 50M per year for new force structure

NOTE: O&S = operations and support.

[a] U.S. Department of Defense, *Fiscal Year (FY) 2017 President's Budget Submission: Army Justification Book of Research, Development, Test & Evaluation, Army RDT&E*, Vol. 2: *Budget Activity 5*, Washington, D.C., February 2016b.

1. Procure MOTS Foreign Design

There are several existing ASMs on the market. For our purposes, the Saab RBS-15 and the Konigsberg Naval Strike Missile are good examples. They each have a truck-mounted version that carries four missile canisters. Nonrecurring costs would be incurred for initial training and spares and to establish the documentation and training environment for continued operation. There may be a cost for intellectual property, and the cost and availability of spares are not as clear-cut as for U.S. systems. Based on recent Polish purchases of a squadron of Naval Strike Missile ground-based missile launchers for $150 million, we assume that a U.S. fires battalion would require $200 million to $300 million, including C2, sensors, and other equipment.[30]

Another viable alternative is that the United States purchases the equipment for an RBS-15 or Naval Strike Missile battalion. The purchased equipment could potentially use the existing HIMARS force structure, either by converting a HIMARS battalion to operate these MOTS systems or by adding another HIMARS-like battalion to the Army's force structure. One RBS-15 or Naval Strike Missile battalion would require 16 launchers and 64 initial rounds. Launchers should cost about $6 million, given that a HIMARS launcher costs $5.8 million,[31] and we assume the RBS-15 or Naval Strike Missile would cost about $0.7 million, based on the $0.9 million cost of the Joint Air-to-Surface Standoff Missile (JASSM) baseline.[32] This results in a total initial procurement cost of $340 million.

If HIMARS battalions are converted to the ASM mission, there should be no increase in O&S cost. If new battalions are added, the cost would be $35 million per year per battalion, including personnel costs. There might be an extra cost for training missiles, depending on how the training approach relates to HIMARS. Here, we assume that a two-battery battalion would be added, with one battery available

---

[30] *Naval Today,* "Poland Orders NSM from Kongsberg," December 19, 2014.

[31] U.S. Department of Defense, *HIMARS*, Selected Acquisition Report, Washington, D.C., December 2011.

[32] U.S. Department of Defense, *Joint Air-to-Surface Standoff Missile (JASSM)*, Selected Acquisition Report, Washington, D.C., December 2015.

for deployments or exercises while the other battery conducts home-station training.

2. Adapt ATACMS

We have selected HIMARS as the launcher for our cost estimate of employing the ASM variant of the ATACMS. It may also be possible to use the armored M270 MLRS in this role. A HIMARS battalion has 16 launchers, each carrying one ATACMS missile. The battalion has 241 personnel and an annual operations and support (O&S) cost on the order of $35 million,[33] for a battalion based in the continental United States. The incremental costs for deploying overseas are $15 million per year.

The ATACMS adaptation would be one of many munitions for the HIMARS. One HIMARS battalion would require 16 ATACMS, plus reloads. With 1.5 reloads, the initial procurement would be 40 missiles, but we elected to procure 160 missiles in our estimate to service the same number of ship targets as the MOTS option costed above. Procurement costs would therefore be $140 million. There would be no negative effect to force structure, because meeting the ASM mission would be a matter of selecting the correct munition.

GPS/INS guidance is used for existing GMLRS. For anti-ship use, a terminal homing capability is required. We assume that an infrared seeker would provide this capability and require modification to the missile C2 hardware and software. Physical integration of the seeker would not pose extensive weight penalties and should be straightforward. In 2012, the ATACMS Alternative Warhead Program was estimated to cost on the order of $1 billion to develop three alternative warheads, including a unitary warhead. The average unit procurement cost was estimated to be $130,000.[34] For the unitary warhead, there were 4.5 years from the start of engineering, manufacturing, and development to full-rate production.

---

[33] Derived from the U.S. Army's Force and Organization Cost Estimating System (FORCES), January 12, 2014.

[34] U.S. Department of Defense, *GMLTS/GMLRS AW*, Selected Acquisition Report, Washington, D.C., December 2012c.

The development phase would require numerous launches for effectiveness and reliability development. We estimate the development for one warhead at $400 million, with an average unit procurement cost of $150,000, in addition to the $720,000 price for a basic ATACMS missile.[35]

3. Adapt LRASM

LRASM is to be a stealthy anti-ship cruise missile. The Defense Advanced Research Projects Agency and the U.S. Navy started the program in 2009. Lockheed Martin is developing it based on JASSM, with air- and ship-launched versions planned.[36] To adapt the ship-launched version (once completed) to ground launch and to provide for particular Army requirements, we estimate an additional development cost range of $200 million to $300 million. *Defense Industry Daily* described a sophisticated set of requirements:

> The US military is also expecting an environment where enemies try to jam or destroy the GPS system and encrypted datalink transmissions, compounding its difficulties in targeting opponents if it can't get many of its platforms through advanced air defenses. Those considerations underline the importance of autonomous targeting. Beyond their anti-jamming digital GPS, therefore, LRASM will also rely on a 2-way data link, a radar sensor that can detect ships (and might also be usable for navigation), and a day/night camera for positive identification and final targeting.[37]

In consideration of the above, we chose a $2 million unit cost, based on the $1.25 million cost of the JASSM–Extended Range (JASSM-ER).[38]

Based on the JASSM and with the solid rocket booster required for ground operation, LRASM will be several feet longer than the

---

[35] U.S. Army, *Procurement Programs: Committee Staff Procurement Backup Book; Fiscal Year (FY) 2010 Budget Estimates, Missile Procurement*, Washington D.C., May 2009.

[36] *Defense Industry Daily*, 2014.

[37] *Defense Industry Daily*, 2014.

[38] U.S. Department of Defense, 2015.

GMLRS or RBS-15. Assuming that the Navy develops LRASM for vertical launching system deployment on ships, the Army would have to develop a specialized ground launcher. The HIMARS launcher was developed for $306 million in a three-year program. The launcher directly implemented a specific existing Army rocket. To address a new munition, we expect a launcher-development cost on the order of $500 million. The unit production cost should be greater than HIMARS, on the order of $7 million, and should carry four LRASMs.

One battalion of 16 launchers each would cost $112 million. An initial ammunition set and four reloads would require 160 missiles, or $320 million. O&S costs would follow the model of the RBS-15 Naval Strike Missile, not including any increment for conversion or $25 million for deploying one battalion, plus training missile consumption approach. Timing for a ground-based version of the LRASM depends on LRASM progress.

4. Field IFPC-2-I to Improve Short-Range Air Defenses

The IFPC-2-I is a mobile, ground-based weapon system that will be designed to acquire, track, engage, and defeat UASs; cruise missiles; and rockets, artillery, and mortars (RAM). The system will provide 360-degree protection and will simultaneously engage threats arriving from different azimuths. A block acquisition approach will be used to provide this capability. The block 1 system will consist of one or more interceptors, the development of technical fire control and a Multi-Mission Launcher, and a Sentinel software upgrade to support the counter-UAS and cruise missile missions. The block 2 system will develop interceptors, sensors, and technical fire control to support the counter-RAM mission. The IFPC-2-I system will use the IAMD C2 open system architecture. This architecture includes multiple sensors with the capability to provide target track data through the IAMD Engagement Operations Center to the IFPC-2-I system. The IFPC-2-I system will be transportable by Army common mobile platforms. The IFPC-2-I product entered technology maturation and risk reduction in fiscal year 2014.[39]

---

[39] U.S. Army, undated.

A rough order-of-magnitude analysis of the cost to develop and field the IFPC-2-I would include $607 million for development.[40] When the IFPC-2-I is fully deployed, there will be two active-duty and seven reserve battalions. Each battalion will have three batteries of three platoons each. The estimated cost per platoon—based on four launch vehicles with 15 missiles each, one Sentinel radar, associated C2 hardware, and essential ground support equipment—is approximately $105 million, and the estimated cost per battalion is $950 million. This includes acquiring 1,350 AIM-9 missiles to provide an initial load and 1.5 reloads per battalion.

Assuming 58 active-duty billets per platoon, plus the annual maintenance material, the estimated annual O&S cost per active-duty battalion with 522 personnel is about $51 million.

5. Develop an SSM Pershing II–Like Ballistic Missile

A reasonable analog to developing a new Pershing II–like ballistic missile is the Trident II, because both are solid-fueled accurate missiles. Current production cost for the Trident is $72 million. Solid rocket motors make up most of the mass of these missiles, but the electronics drive the costs. A good working number for each missile—given the smaller size, range, and number of warheads for the ground-based missile—is in the range of $25 million to $35 million, plus a transporter-erector-launch vehicle in the $5 million to $10 million range. Stephen Schwartz, in *Atomic Audit*, lists a cost of $1,500 million for Pershing II research, development, test, and evaluation (RDT&E) from 1977 through 1984. This would translate to $3.729 billion, deflating RDT&E dollars from 1980 to 2017, although development costs for this program might be higher given that the Pershing II program was a follow-on to Pershing I.[41] Therefore, we use an estimate of $4.5 billion for the development costs of this program. One battalion of 16

---

[40] U.S. Department of Defense, *Fiscal Year (FY) 2017 President's Budget Submission: Chemical and Biological Defense Program*, Vol. 4: *Research, Development, Test & Evaluation, Defense-Wide*, Vol. 4, Washington, D.C., February 2016a, p. 389.

[41] Stephen I. Schwartz, *Atomic Audit: The Costs and Consequences of U.S. Nuclear Weapons Since 1940*, Washington, D.C.: Brookings Institution Press, June 1, 1998.

missiles and 1.5 rounds of reloads would be $1 billion to $1.4 billion. New unit O&S costs would be $35 million per year per new battalion, but there would be increased costs for parts. The start of engineering, manufacturing, and development to full-rate production should take on the order of ten years.

Near-Term Force Structure Options for Ground-Based Multi-Domain Fires

In this section, we examine alternative options for deploying a multi-domain fires battalion in the Asia-Pacific or European theater. This multi-domain fires battalion concept would include surface-to-air, surface-to-surface, and surface-to-ship missile capabilities. We begin with options that minimize the new investment needed to put an initial capability into the field and then progressively add features and corresponding costs. For the long-range anti-ship and surface-to-surface missions, we assume that the identification, tracking, and targeting are provided by U.S. Navy or U.S. Air Force ISR systems. Table 7.7 summarizes the options.

Option 1: Deploy Combined Battalion with Existing U.S. and Allied Units and Missiles

The minimum investment option would be to deploy a multi-domain fires battalion based on existing U.S. and allied capabilities and units. An IFPC-2 battery consisting of 12 firing units, three Sentinel radars, and three C2 sets would be assigned from the existing air-defense artillery force. The IFPC-2 units are just now being formed; the battery assigned to the multi-domain fires battalion would take this unit from the U.S. Army air defense force pool as it is being modernized. If minimizing initial investment costs to the Army is a priority, AIM-9 missiles can be assigned from the PACOM or EUCOM theater weapons stocks to reduce the immediate acquisition costs.

Similarly, a HIMARS or MLRS battery with launchers and fire control center could be assigned from the existing U.S. Army rocket artillery force. This battery could employ both Short-range MLRS rock-

Table 7.7
Combined Multi-Domain Fires Battalion Options

Force Structure Option	Anti-Ship Battery	Air Defense Battery	Surface-to-Surface Battery	Long-Range ISR and Targeting	Incremental Costs ($M)
Option 1: combined battalion	Provided by allies	Planned IFPC-2 using joint AIM-9	Existing HIMARS with ATACMS/GMLRS	U.S. Air Force and U.S. Navy	RDT&E: $0 Procurement: $0 O&S: $10M
Option 2: U.S.-owned allied systems	U.S. purchases MOTS	Planned IFPC-2	Existing HIMARS	U.S. Air Force and U.S. Navy	RDT&E: $0 Procurement: $340M O&S: $30M
Option 3: U.S. develops ASM	Anti-ship ATACMS	Planned IFPC-2	Existing HIMARS	U.S. Air Force and U.S. Navy	RDT&E: $400M Procurement: $140M O&S: $15M
Option 1, 2, or 3 with new A2/SSM purchases		Add 450 new AIM-9X: $190M	Add 80 new ATACMS: $60M	U.S. Air Force and U.S. Navy	Add: $250M per deployed multi-domain battalion

NOTE: RDT&E = research, development, test, and evaluation.

ets and ATACMS for attacking surface targets at long range. Weapons would be assigned from Army or theater global stocks. When the LRPF missile is fielded, it could be included in the battery for even longer-range targets.

For the anti-ship role, an initial capability could be established by building a combined battalion incorporating existing ASM batteries operated by selected allies. For Europe, Poland has ground-launched ASM batteries that could be included. Norway, Denmark, and Sweden have fielded ground-launched ASM batteries in the past and could provide a battery to a combined unit if they choose to reestablish such units. In the Pacific, Japan could provide ground-based ASM batteries for combined forces assigned to defend Japan. Additional allies could seek to join as they develop the requisite capabilities.

The long-range SSMs and ASMs included in this option would need ISR capabilities to detect, track, and target enemy forces. As noted, U.S. Air Force, U.S. Navy, and allied airborne sensors could be used to provide these capabilities. The principal advantage of this option is that it minimizes the new costs needed to field a multi-domain fires capability. Already-programmed U.S. Army SAM batteries could be employed using munitions from the global force pool. Existing U.S. Army SSM batteries would be employed with their current missile stocks. New capabilities—specifically, ASM capabilities—would be provided by the nations that U.S. forces have been deployed to defend.

New costs of this option to the U.S. Army would be minimal, consisting only of the incremental O&S costs to operate a battalion on deployments. We estimate these costs to be similar to the incremental cost of deploying a fires battalion overseas from stateside, or about $15 million per year.[42] Since the United States would be operating only two of the three batteries in this option, we estimate annual incremental costs of $10 million to deploy the multi-domain battalion.

There are also important opportunity costs for the Army and the joint force. The IFPC-2 and HIMARS batteries deployed in this battalion would not be available for their previous roles in maneuver or fires units, which might leave an important gap that must be filled. Additionally, the munitions that the IFPC-2 battery would draw from global munition stocks represent a limited resource that would not be available for use by other air, sea, or ground units. If the Army had to pay for the 450 AIM-9 missiles needed to arm the IFPC-2 battery, the procurement cost would increase by $190 million. Buying the 80 ATACMS needed for the surface-to-surface fires battery would increase procurement costs by about $60 million, for a total incremental procurement cost of $250 million.

Option 2: Build U.S. ASM Units with Allied Systems
A more significant investment would be needed to deploy an all-U.S. multi-domain fires unit. One option would be to purchase ASMs, launchers, and any specialized C2 equipment from an allied nation to equip an existing fires battalion. As described above, an anti-ship battalion could be equipped with 16 quad-launchers for the Swedish

---

[42] The FORCES cost model shows an annual O&S cost of an Avenger/IFPC battalion as $56.8 million in the continental United States and $72.0 million in EUCOM.

RBS-15 or Norwegian Naval Strike Missile organized into two batteries of eight quad launchers each. The cost of 16 quad launchers and an initial munitions load of 160 missiles would be about $340 million. The two batteries purchased could alternate between training at home station and deploying for exercises.

A disadvantage of this option would be the added logistical expense of supplying and maintaining equipment unique to this role and not already in the inventory. A further disadvantage would be the need to train crews to operate this unique equipment. We estimate that these disadvantages would double the incremental annual O&S costs.

In this option, IFPC-2 would be used for short-range air defense and ATACMS would be used for surface-to-surface fires. As in option 1, the costs of this option would increase by $250 million if new AIM-9 and ATACMS missiles had to be purchased.

Option 3: Build U.S. ASM Units with U.S.-Built Systems

Alternatively, in this option, the Army could purchase the anti-ship version of the ATACMS currently under development by the DoD Strategic Capabilities Office[43] and convert a HIMARS battalion to fire the anti-ship variant of the ATACMS. If no changes need to be made to the basic HIMARS equipment, a given battery could select between firing anti-ship ATACMS, surface-to-surface ATACMS, and GMLRS rockets. A battalion of two batteries would be outfitted with anti-ship ATACMS to enable one unit to train while the other conducted exercises. As in the other two options, IFPC-2 would be used for short-range air defense, and ATACMS would be used for surface-to-surface fires. As stated earlier, we estimate the cost of this option to be $400 million in development costs and $140 million in procurement costs, with an additional procurement cost of $250 million if new AIM-9 and ATACMS missiles have to be purchased.

An advantage of this option is that the multi-domain fires battalion would use a HIMARS battery for each of the anti-ship and surface-to-surface fires missions. If the ASMs and SSMs proved to be

---

[43] Aaron Mehta, "Anti-Naval ATACMS, 'Big' Swarming Breakthroughs from Strategic Capabilities Office," *Defense News*, October 28, 2016.

interchangeable with the launchers, this would greatly increase the flexibility of the battalion, thus allowing it to essentially double down on either mission as targeting needs varied. It would also add to an adversary's uncertainty about the capabilities of any HIMARS unit, since it would be impossible to tell which unit had the ASMs.

Future Force Structure Options for Ground-Based Multi-Domain Fires

In the longer run, additional multi-domain fires battalions could be outfitted along the lines of any of the options listed above. These battalions could represent a new force structure or modify existing HIMARS or MLRS battalions. In addition, future battalions could be equipped with LRASM or longer ranged surface-to-surface fires.

CHAPTER EIGHT
Recommendations, Open Questions, and Next Steps

A fundamental choice for the United States is whether to focus on building and supporting allied defensive concepts or employing primarily U.S.-only concepts on behalf of allies. That is, should the United States act with and in support of its allies or should it act for them? Ground-based concepts offer a relatively affordable way for allies and partners to take the primary responsibility for their own defense, and they offer an attractive way for the United States to build capabilities and capacity to operate in support of allies.

Based on the findings and conclusions above, we suggest that the Army organize and field a prototype multi-domain fires battalion to develop, test, and exercise joint and combined defensive concepts discussed in the report. Table 8.1 provides a set of recommendations, which are discussed in more detail below.

To start as quickly and inexpensively as possible, long-range ISR and targeting capabilities should be provided by existing U.S. Navy, U.S. Air Force, and allied systems. An existing HIMARS battery should be assigned to provide surface-to-surface fires. These fires could contribute to SEAD missions, disrupt operations at enemy air bases, and attack amphibious forces that have landed on friendly territory. This battery could be equipped with both ATACMS and GMLRS. Improved long-range missiles, such as the 499-kilometer LRPF system, could extend the range of this battery when fielded. If an anti-ship version of the ATACMS or GMLRS is developed, then the anti-ship and surface-to-surface batteries could be interchangeable and assigned

Table 8.1
Recommendations for Multi-Domain Fires Battalion

Multi-Domain Capability	Recommendations for the U.S. Army
Long-range ISR and targeting	• Rely on U.S. Navy, U.S. Air Force, and allied forces and systems
Surface-to-surface capabilities	• Assign HIMARS battalion equipped with ATACMS and GMLRS

Anti-ship capabilities	• Assign ASM battery
	  – Begin with combined U.S.-allied batteries in multi-domain battalions
	  – Develop anti-ship version of ATACMS for U.S. HIMARS battery
	  – Develop C2 systems and procedures, as well as fire support coordination measures with maritime components

Air and cruise missile defense capabilities	• Develop deployable IFPC-2 minimum engagement packages (MEPs) • Assign battery of three IFPC-2 MEPs into multi-domain battalion

targets more flexibly—perhaps even doubling the firepower that could be delivered against either target set as operational priorities dictate.

For the anti-ship role, an initial capability could be established by building a combined battalion incorporating existing ASM batteries operated by selected allies, such as Poland and Japan. Additional allies could seek to join as they develop the requisite capabilities. If current development programs succeed in building versions of the ATACMS or GMLRS with a terminal guidance package for anti-ship operations, the multi-domain battalion should incorporate a U.S. Army anti-ship battery capable of operating them. We would prefer this as the longer-term option to provide more flexibility for employing both ground-attack and anti-ship versions of the ATACMS from the same launchers. Using systems already in the U.S. inventory should also simplify logistics and support operations.

Finally, a short-range air and cruise missile defense battery could be assigned from the forces being formed to operate IFPC-2. To provide a capability that can operate in small numbers, the U.S. Army should develop and deploy minimum engagement packages (MEPs) for exercises and demonstrations with allies and partners. Three such platoon-sized MEPs could be assigned as the air and cruise missile defense battery.

If the prototype battalion proves useful, additional multi-domain fires battalions could also be formed. These battalions could represent new force structure or assignments of existing HIMARS and MLRS batteries and IFPC-2 platoons. In addition, future battalions could be equipped with LRASMs or longer ranged surface-to-surface fires.

After an initial set of joint operating concepts have been developed, the Army should work with key allies and partners to build combined concepts and tactics, techniques, and procedures. Allies and partners could include Japan, the Philippines, and Taiwan in the Asia-Pacific region; NATO nations (Poland, for example, already fields Norway's Naval Strike Missile in coastal defense squadrons); Sweden; Finland; and Gulf Cooperation Council members in the Persian Gulf region. The United States could plan and host joint and combined exercise events to refine these concepts and provide continuing training for allies and partners.

Areas for Further Analysis and Development

This analysis leaves open several questions that warrant further analysis and development.

Force Size and Mix

Our recommendations for a prototype multi-domain fires battalion are designed to enable joint concept development and experimentation. Therefore, the mix of surface-to-surface, anti-ship, and air and cruise missile defense capabilities is balanced to enable experimentation rather than meet particular operational objectives. In practice, the size and mix of such capabilities will vary based on the objectives and missions in a particular scenario. A deeper examination of the size and mix of the multi-domain fires battalion warrants further investigation. In addition, the large "advise and assist" role of the multi-domain fires battalion needs to be considered, because that will require more officers, senior enlisted, and communications capability than normal HIMARS and MLRS units.

Mission Scope
This work has focused primarily on surface-to-surface, anti-ship, and surface-to-air capabilities without examining potential missions for a multi-domain task force in the space and cyber domains or multi-domain battle roles that employ maneuver, force protection, aviation, or sustainment functions. Additionally, other multi-domain concepts of operation are worthy of consideration, such as potentially establishing land-based defense of carrier battle groups, defending land-bases from asymmetric mortar or SOF attacks, among others. An examination of the broader mission scope of a multi-domain battle task force is an important area for future work.

Joint Wargaming
We analyzed the potential demands associated with allied and U.S. ground forces organized and equipped with multi-domain fires. One way to test the value proposition of such a unit is through a series of formally adjudicated wargames where active Blue and Red teams are afforded the choice of employing or countering the proposed multi-domain fires battalion. Such wargames would be a useful analytic component to the proposed joint concept development and experimentation and would enable further refinement of these ideas.[1]

Fire Support Coordination
The Army does have multi-domain coordination experience in air and missile defense, but coordination of fires between the Joint Force Land Component Commander and the Joint Force Maritime Component Commander is new. This will require new doctrine and new coordination measures with the maritime components.

---

[1] For a discussion of a detailed wargame, built on the basis of decades of rigorous quantitative and qualitative analyses, see Shlapak and Johnson, 2016.

Figures and Tables

Figures

6.1. Notional Ground-Based ASM Operational Capabilities........ 78
6.2. Notional Capabilities to Enhance Survivability of Ground-Based Defenses... 81
6.3. Maritime Area Coverage from Yonaguni and Iriomote......... 83
6.4. Yonaguni and Iriomote Area and Population 84
6.5. Ranges from the Chinese Mainland to the Philippines........ 86
6.6. Coverage of the South China Sea from the Philippines........ 87
6.7. Coverage of Maritime Approaches to Taiwan.................... 89
6.8. Coverage of the Baltic Sea from Russia and NATO Nations.. 93
6.9. Coverage of the Baltics, Belarus, and Russia from Poland and Estonian Coastal Islands... 94
6.10. Coverage of the Black Sea and Bosphorus Strait from NATO Nations 95
6.11. Coverage of the Persian Gulf... 96
6.12. Coverage of Iranian Bases from the Arabian Peninsula........ 97
7.1. Map of the PACOM AOR.. 100
7.2. Chinese Missile Coverage of Western Pacific.................... 105
7.3. Potential Basing Locations Within 300 Kilometers of Two Flash Points... 111
7.4. Potential Basing Locations Within 500 Kilometers of Two Flash Points... 112
7.5. Depth of Attack and Responsiveness of Strike Assets......... 120
7.6. Response Time and Coverage of Airborne Versus Land-Based Strike Assets 121
7.7. HIMARS Raid Concept... 127

Tables

S.1. Ground-Based A2/AD Missions—Findings and Conclusions xvi
S.2. Recommendations for Prototype Multi-Domain Fires Battalion for Joint Concept Development and Experimentation xviii
2.1. China's Core Interests 16
6.1. Chinese Long-Range Missile Systems 76
6.2. Exemplar Allied ASM Systems and U.S. Air, Sea, and Ground Missiles 80
6.3. Russian Long-Range Missile Systems 92
7.1. Approximate Distances Between Representative Locations in the PACOM AOR 102
7.2. Net Effect of Geography, Arms Control Agreements, Policy, and Current Weapon Systems 103
7.3. Existing Joint Anti-Ship Delivery Capacity 109
7.4. Current and Projected Chinese Naval Forces 109
7.5. Ground Based A2/AD Missions—Findings and Conclusions 135
7.6. Estimated Costs for Initial Deployed Battalion 137
7.7. Combined Multi-Domain Fires Battalion Options 144
8.1. Recommendations for Multi-Domain Fires Battalion 150

Abbreviations

A2/AD	anti-access/area denial
AOR	area of responsibility
APC	armored personnel carrier
ASB	Air-Sea Battle
ASM	anti-ship missile
ATACMS	Army Tactical Missile System
C2	command and control
COIN	counterinsurgency
DoD	U.S. Department of Defense
EDCA	Enhanced Defense Cooperation Agreement
EEZ	exclusive economic zone
EUCOM	U.S. European Command
FORCES	Force and Organization Cost Estimating System
GDP	gross domestic product
GMLRS	Guided Multiple Launch Rocket System
GPS	Global Positioning System
GSDF	Ground Self-Defense Forces
HIMARS	High-Mobility Artillery Rocket System
HI-RAIN	HIMARS Rapid Infiltration
IAMD	Integrated Air and Missile Defense

IBCS	Integrated Air and Missile Defense Battle Command System
IFPC-2	Indirect Fire Protection Capability–Increment 2
IFPC-2-I	Indirect Fire Protection Capability–Increment 2–Intercept
INF	Intermediate-Range Nuclear Forces
INS	inertial navigation system
ISR	intelligence, surveillance, and reconnaissance
JASSM	Joint Air-to-Surface Standoff Missile
JASSM-ER	Joint Air-to-Surface Standoff Missile–Extended Range
JOAC	Joint Operational Access Concept
LRASM	Long-Range Anti-Ship Missile
LRPF	Long-Range Precision Fires
MEP	minimum engagement package
MLRS	multiple launch rocket system
MND	Ministry of National Defense
MOTS	military-off-the-shelf
NATO	North Atlantic Treaty Organization
NSS	National Security Strategy
O&S	operations and support
OEF-P	Operation Enduring Freedom–Philippines
PACOM	U.S. Pacific Command
PLA	People's Liberation Army
PLAAF	People's Liberation Army Air Force
PLAN	People's Liberation Army Navy
RBS	Robotsystem
ROE	rules of engagement
SAM	surface-to-air missile
SDF	Self-Defense Forces

SEAD	suppression of enemy air defenses
SOF	special operations forces
SSM	surface-to-surface missile
TRA	Taiwan Relations Act
UAS	unmanned aerial system
UAV	unmanned aerial vehicle

References

10th National People's Congress and Chinese People's Political Consultative Conference, 3rd session, Anti-Secession Law, Beijing, March 14, 2005.

32nd Army Air and Missile Defense Command, *Operation Iraqi Freedom Theater Air and Missile Defense History*, Fort Bliss, Tex., September 2003.

62nd Airlift Wing Public Affairs, "Airmen, Soldiers Train to Rapidly Infiltrate Enemy Territory," Air Mobility Command, June 28, 2016. As of February 15, 2017:
http://www.amc.af.mil/News/Article-Display/Article/818200/airmen-soldiers-train-to-rapidly-infiltrate-enemy-territory/

Agence France-Presse, "President Says Philippines to Spend $1.8 Billion on Military Modernization," *Defense News*, December 21, 2015. As of November 30, 2016:
http://www.defensenews.com/story/defense/2015/12/21/president-says-philippines-spend-military-modernization/77726792/

Air-Sea Battle Office, *AIR-SEA BATTLE: Service Collaboration to Address Anti-Access & Area Denial Challenges*, Washington, D.C.: U.S. Department of Defense, May 2013. As of November 23, 2016:
http://archive.defense.gov/pubs/ASB-ConceptImplementation-Summary-May-2013.pdf

Apthorp, Claire, "Light Armoured Vehicle Procurement," *Defence Review Asia*, October 26, 2011. As of November 30, 2016:
http://www.defencereviewasia.com/articles/133/Light-armoured-vehicle-procurement-in-Asia

Associated Press, "Philippine President Duterte Announces Separation from U.S.," October 20, 2016.

Balle, Joakim Kasper Oestergaard, "MIM-104F Patriot PAC-3: About Patriot and PAC-3," BGA-Aeroweb, September 6, 2016. As of November 30, 2016:
http://www.bga-aeroweb.com/Defense/Patriot-PAC-3.html

BBC News, "China Establishes Air Defense Zone over East China Sea," November 23, 2013. As of November 29, 2016:
http://www.bbc.com/news/world-asia-25062525

Bi Xinglin, ed., *Campaign Theory Study Guide*, Beijing: National Defense University Press, 2002.

Bonds, Timothy M., Eric V. Larson, Derek Eaton, and Richard E. Darilek, *Strategy-Policy Mismatch: How the U.S. Army Can Help Close Gaps in Countering Weapons of Mass Destruction*, Santa Monica, Calif.: RAND Corporation, RR-541-RC, 2014. As of December 15, 2016:
http://www.rand.org/pubs/research_reports/RR541.html

Bush, Richard, *Uncharted Strait: The Future of China-Taiwan Relations*, Washington, D.C.: Brookings Institution Press, 2013.

Chase, Michael S., "China's Search for a 'New Type Great Power Relationship,'" *China Brief*, Vol. 12, No. 17, September 7, 2012. As of November 29, 2016:
http://www.jamestown.org/single/
?tx_ttnews%5Btt_news%5D=39820&no_cache=1#.VJL3d7hxQA

Chase, Michael, Jeffrey Engstrom, Tai Ming Cheung, Kristen Gunness, Scott Warren Harold, Susan Puska, and Samuel K. Berkowitz, *China's Incomplete Military Transformation: Assessing the Weaknesses of the People's Liberation Army (PLA)*, Santa Monica, Calif.: RAND Corporation, RR-893-USCC, 2015. As of February 24, 2017:
http://www.rand.org/pubs/research_reports/RR893.html

China Military Online, "Chinese Navy to Conduct Drills Around Xisha Islands," July 6, 2016. As of November 29, 2016:
http://english.chinamil.com.cn/news-channels/2016-07/06/content_7138794.htm

Cliff, Roger, Mark Burles, Michael Chase, Derek Eaton, and Kevin L. Pollpeter, *Entering the Dragon's Lair: Chinese Antiaccess Strategies and Their Implications for the United States*, Santa Monica, Calif.: RAND Corporation, MG-524-AF, 2007. As of January 24, 2017:
http://www.rand.org/pubs/monographs/MG524.html

Cliff, Roger, John F. Fei, Jeff Hagen, Elizabeth Hague, Eric Heginbotham, and John Stillion, *Shaking the Heavens and Splitting the Earth: Chinese Air Force Employment Concepts in the 21st Century*, Santa Monica, Calif.: RAND Corporation, MG-915-AF, 2011. As of January 24, 2017:
http://www.rand.org/pubs/monographs/MG915.html

Colaresi, Michael P., Karen Rasler, and William R. Thompson, *Strategic Rivalries in World Politics: Position, Space, and Conflict Escalation*, Cambridge, UK: Cambridge University Press, 2008.

Congress of the Philippines, An Act Amending Republic Act No. 7898. Establishing the Revised AFP Modernization Program and for Other Purposes (Armed Forces of the Philippines New Modernization Act), Republic Act No. 10349, Manila, 15th Congress, 3rd regular session, December 11, 2012.

Corr, Anders, "Chinese Bomber Buzzes Philippines' Scarborough Shoal in Latest Salvo of U.S.-China Signaling War," *Forbes*, July 17, 2016.

Council on Foreign Relations, *China's Maritime Disputes*, Washington, D.C., 2016. As of November 30, 2016:
http://www.cfr.org/asia-and-pacific/chinas-maritime-disputes/p31345#!/p31345

Cudis, Christine C., "AFP Budget to Be Spent on Modernization of External Defenses," *PhilStar Global*, May 27, 2016. As of November 30, 2016:
http://www.philstar.com/headlines/2016/05/27/1587520/
afp-budget-be-spent-modernization-external-defense-military-official

Cutler, Thomas J., *The Battle for the Paracel Islands*, Annapolis, Md.: Naval Institute Press, 1974.

Dai Bingguo, "Adhere to the Path of Peaceful Development," Ministry of Foreign Affairs, People's Republic of China, December 6, 2010.

Defense Industry Daily, "Kongsberg's NSM/JSM Anti-Ship and Strike Missile Attempts to Fit in Small F-35 Stealth Bay," November 12, 2015. As of December 1, 2016:
http://www.defenseindustrydaily.com/
norwegian-contract-launches-nsm-missile-03417/

———, "LRASM Missiles: Reaching for a Long-Range Punch," May 17, 2016. As of December 1, 2016:
http://www.defenseindustrydaily.com/
lrasm-missiles-reaching-for-a-long-reach-punch-06752/

Defense News, "Interview: Nien-Dzu Yang, Taiwan's Vice Minister of Defense—Policy," November 12, 2012.

———, "Report: China's UAVs Could Challenge Western Domination," June 25, 2013.

———, "Philippines Hikes Defense Budget 25%," July 21, 2015. As of November 30, 2016:
http://www.defensenews.com/story/defense/policy-budget/budget/2015/07/21/
philippines-hikes-defense-budget-25-percent-amid-south-China-sea-dispute--China/
30464145/

Denyer, Simon, "Philippine Leader Duterte Now Wants U.S. Troops Out 'in the Next Two Years,'" *Washington Post*, October 26, 2016. As of February 11, 2017: https://www.washingtonpost.com/world/philippines-duterte-now-wants-us-troops-out-in-two-years/2016/10/26/
32bec8a5-85
84-4d95-8e9d-4d7762865055_story.html?utm_term=.4dd7bfc4ecce

Edwards, Jane, "Reports: HASC's 2017 NDAA Calls for Army to Replace Patriot Radar, Assess Land-Based Anti-Ship Missiles," *ExecutiveGov*, April 27, 2016. As of November 23, 2016: http://www.executivegov.com/2016/04/reports-hascs-2017-ndaa-calls-for-army-to-replace-patriot-radar-assess-land-based-anti-ship-missiles/

Federation of American Scientists, Annex G, "APP 7—Artillery Raid, Tactics, Techniques, and Procedures for Air Assault Operations," March 17, 1999.

Fonbuena, Carmela, "PH Finalizing Air Defense Radar Deal with Israel," *Rappler*, July 9, 2014. As of November 30, 2016: http://www.rappler.com/nation/62841-philippines-air-defense-radar

Fontenot, Gregory, E. J. Degen, and Dave Tohn, *On Point—The United States Army in Operation Iraqi Freedom*, Fort Leavenworth, Kan.: Combat Studies Institute Press, 2004. As of November 23, 2016: http://usacac.army.mil/cac2/cgsc/carl/download/csipubs/OnPointI.pdf

Foss, Christopher F., and James C. O'Halloran, *IHS Jane's Land Warfare Platforms: Artillery & Air Defence*, London: Jane's Information Group, 2015.

Freedberg, Sydney J., Jr., "Army Should Build Ship-Killer Missiles: Rep. Randy Forbes," *Breaking Defense*, October 12, 2014. As of November 23, 2016: http://breakingdefense.com/2014/10/
army-should-build-ship-killer-missiles-rep-randy-forbes/

———, "New Army Long-Range Missile Might Kill Ships, Too: LRPF," *Breaking Defense*, October 13, 2016. As of December 1, 2016: http://breakingdefense.com/2016/10/
new-army-long-range-missile-might-kill-ships-too-lrpf/

Fuller, Malcolm, and David Ewing, *IHS Jane's Weapons: Naval*, London, Jane's Information Group, 2013.

Gerts, Bill, "Pacific Command: China Seeking to Control South China Sea; Harris Calls for Buildup of U.S. Missiles, Weapons in Region," *Washington Free Beacon*, February 24, 2016. As of November 23, 2016: http://freebeacon.com/national-security/
pacific-command-china-seeking-to-control-south-china-sea/

Gill, Bates, *Rising Star: China's New Security Diplomacy*, Washington, D.C.: Brookings Institute Press, 2010.

Global Security, "Chinese Warships," web page, undated. As of March 1, 2017:
http://www.globalsecurity.org/military/world/china/navy.htm

———, "Operation Enduring Freedom—Philippines," May 7, 2011. November 30, 2016:
http://www.globalsecurity.org/military/ops/enduring-freedom-philippines.htm

———, "Exercise Balikatan: Shouldering the Load Together," October 4, 2016. As of November 30, 2016:
http://www.globalsecurity.org/military/ops/balikatan.htm

Global Times, "Power Game Decides Post-Arbitration Order," July 5, 2016. As of November 30, 2016:
http://www.globaltimes.cn/content/992320.shtml

Hagen, Jeff, Forrest E. Morgan, Jacob Heim, and Matthew Carroll, *The Foundations of Operational Resilience—Assessing the Ability to Operate in an Anti-Access/Area Denial (A2/AD) Environment: The Analytical Framework, Lexicon, and Characteristics of the Operational Resilience Analysis Model (ORAM)*, Santa Monica, Calif.: RAND Corporation, RR-1265-AF, 2016. As of January 24, 2017:
http://www.rand.org/pubs/research_reports/RR1265.html

Harris, Harry B., Jr., "Role of Land Forces in Ensuring Access to Shared Domains," speech presented at the Institute of Land Warfare LANPAC Symposium, Waikiki, May 25, 2016. As of November 23, 2016:
http://www.pacom.mil/Media/Speeches-Testimony/Article/781889/lanpac-symposium-2016-role-of-land-forces-in-ensuring-access-to-shared-domains/

Heath, Timothy R., "What Does China Want? Discerning the CHINA's National Strategy," *Asian Security*, Vol. 8, No. 1, 2012, pp. 54–72. As of November 23, 2016:
http://www.tandfonline.com/doi/abs/10.1080/14799855.2011.652024

———, "China and the U.S. Alliance System," *The Diplomat*, June 11, 2014. As of November 28, 2016:
http://thediplomat.com/2014/06/China-and-the-u-s-alliance-system/

Heginbotham, Eric, Michael Nixon, Forrest E. Morgan, Jacob L. Heim, Jeff Hagen, Sheng Li, Jeffrey Engstrom, Martin C. Libicki, Paul DeLuca, David A. Shlapak, David R. Frelinger, Burgess Laird, Kyle Brady, and Lyle J. Morris, *The U.S.-China Military Scorecard Forces, Geography, and the Evolving Balance of Power, 1996–2017*, Santa Monica, Calif.: RAND Corporation, RR-392-AF, 2015. As of November 23, 2016:
http://www.rand.org/pubs/research_reports/RR392.html

Hille, Kathrin, "U.S. Seeks to Calm Beijing Containment Fears," *Financial Times*, December 8, 2011.

Hsu, Kimberly, and Craig Murray, *China and International Law in Cyber Space*, Washington, D.C.: U.S.-China Economic and Security Review Commission, May 6, 2014. As of November 29, 2016:
http://origin.www.uscc.gov/sites/default/files/Research/China%20 International%20Law%20in%20Cyberspace.pdf

Huang, Alexander Chieh-cheng, "The United States and Taiwan's Defense Transformation," *Taiwan-U.S. Quarterly Analysis*, Brookings Institute, February 2010. As of November 30, 2016:
http://www.brookings.edu/research/opinions/2010/02/taiwan-defense-huang

Hughes, Robin, "Sweden Reactivates RBS15-Based Mobile Coastal Defence Systems," *IHS Jane's Missiles & Rockets*, December 30, 2016.

ID James, "Israeli Arms Sales to Philippines and Its Implications: The Present Scenario," August 2013.

International Crisis Group, "Stirring Up the South China Sea," *Asia Report*, No. 223, April 23, 2012.

———, "Dangerous Waters: China-Japan Relations on the Rocks," *Asia Report*, No. 245, April 8, 2013. As of November 29, 2016:
https://archive.org/stream/696878-icg-dangerous-waters-china-japan-relations-on/ 696878-icg-dangerous-waters-china-japan-relations-on_djvu.txt

International Institute for Strategic Studies, *The Military Balance 2012*, London, March 7, 2012. As of December 1, 2016:
https://www.iiss.org/en/publications/military%20balance/issues/ the-military-balance-2012-77da

Japan Times, "Chinese Activists Land on Senkaku Islet; Japan Arrests 14," August 16, 2012. As of November 29, 2016:
http://www.japantimes.co.jp/news/2012/08/16/national/ chinese-activists-land-on-senkaku-islet-japan-arrests-14/

Joint Chiefs of Staff, *The National Military Strategy of the United States of America: A Strategy for Today; A Vision for Tomorrow*, Washington, D.C.: Office of the Chairman of the Joint Chiefs of Staff, 2004. As of November 22, 2016:
www.strategicstudiesinstitute.army.mil/pdffiles/nms2004.pdf

———, *The National Military Strategy of the United States of America 2015: The United States Military's Contribution to National Security*, Washington, D.C., June 2015. As of November 23, 2016:
http://www.jcs.mil/Portals/36/Documents/Publications/2015_National_Military_ Strategy.pdf

Joint Publication 3-32, *Command and Control for Joint Maritime Operations*, Washington, D.C.: Joint Chiefs of Staff, August 7, 2013.

Kazianis, Harry, "China's Expanding Cabbage Strategy: After Pursuing a Cabbage Strategy in the South China Sea for Years, Could Beijing Adopt This Against Japan?" *The Diplomat*, October 29, 2013.

Keck, Zachary, "Taiwan Acquires Submarine-Launched Anti-Ship Missiles," *The Diplomat*, December 27, 2013. As of December 1, 2016:
http://thediplomat.com/2013/12/taiwan-acquires-submarine-launched-anti-ship-missiles/

Kelly, Terrence K., Anthony Atler, Todd Nichols, and Lloyd Thrall, *Employing Land-Based Anti-Ship Missiles in the Western Pacific*, Santa Monica, Calif.: RAND Corporation, TR-1321-A, 2013. As of November 23, 2016:
http://www.rand.org/pubs/technical_reports/TR1321.html

Kelly, Terrence K., David C. Gompert, and Duncan Long, *Smarter Power, Stronger Partners, Vol. 1: Exploiting U.S. Advantages to Prevent Aggression*, Santa Monica, Calif.: RAND Corporation, RR-1359-A, 2016. As of November 23, 2016:
http://www.rand.org/pubs/research_reports/RR1359.html

Kopp, Carlo, "PLA Cruise Missiles," Air Power Australia, January 27, 2014. As of December 1, 2016:
http://www.ausairpower.net/APA-PLA-Cruise-Missiles.html#mozTocId838105

Krepinevich, Andrew, Barry Watts, and Robert Work, *Meeting the Anti-Access and Area-Denial Challenge*, Washington, D.C.: Center for Strategic and Budgetary Assessments, 2003.

Lester, Karl, and Clarissa Batino, "Outgunned Philippine General Seeks Arms Upgrade as China Expands," *Bloomberg News*, September 5, 2014.

Lostumbo, Michael, "A New Taiwan Strategy to Adapt to PLA Precision Strike Capabilities," in Roger Cliff, Phillip C. Saunders, and Scott Warren Harold, eds., *New Opportunities and Challenges for Taiwan's Security*, Santa Monica, Calif.: RAND Corporation, CF-279-OSD, 2011, pp. 127–136. As of February 25, 2017:
http://www.rand.org/pubs/conf_proceedings/CF279.html

Lostumbo, Michael, David R. Frelinger, James Williams, and Barry Wilson, *Air Defense Options for Taiwan: An Assessment of Relative Costs and Operational Benefits*, Santa Monica, Calif.: RAND Corporation, RR-1051-OSD, 2016. As of December 15, 2016:
http://www.rand.org/pubs/research_reports/RR1051.html

Lui, Kevin, "Philippine President Rodrigo Duterte Threatens to End Defense Pact with the U.S.," *Time*, October 3, 2016. As of November 30, 2016:
http://time.com/4516231/philippines-rodrigo-duterte-tdefense-agreement-us-edca-drug-war/

Mangosing, Frances, "More Ships, Planes, Combat Gear for AFP in 2016," *Inquirer.net*, January 1, 2016. As of November 30, 2016:
http://newsinfo.inquirer.net/751767/more-ships-planes-combat-gear-for-afp-in-2016-activists-land-on-senkaku-islet-japan-arrests-14/

Manila Standard Today, "PH Plans to Tap Israel for Missile Launchers," June 15, 2013.

Mayger, James, and Yuji Nakamura, "Japan Protests Intrusion of Armed Chinese Vessels into Its Waters," *Bloomberg*, December 25, 2015. As of November 30, 2016:
http://www.bloomberg.com/news/articles/2015-12-26/japan-coast-guard-says-three-chinese-ships-near-senkaku-islands

Medeiros, Evan S., Keith Crane, Eric Heginbotham, Norman D. Levin, Julia F. Lowell, Angel Rabasa, and Somi Seong, *Pacific Currents: The Responses of U.S. Allies and Security Partners in East Asia to China's Rise*, Santa Monica, Calif.: RAND Corporation, MG-736-AF, 2008. As of December 15, 2016:
http://www.rand.org/pubs/monographs/MG736.html

Mehta, Aaron, "Anti-Naval ATACMS, 'Big' Swarming Breakthroughs from Strategic Capabilities Office," Defense News, October 28, 2016.

Military-Today, "WS-2: Multiple Launch Rocket System," web page, undated. As of November 29, 2016:
http://www.military-today.com/artillery/ws2.htm

Ministry of Defense of Japan, *Medium Term Defense Program (FY2014–FY2018)*, Tokyo, December 17, 2013a. As of February 24, 2017:
http://www.mod.go.jp/j/approach/agenda/guideline/
2014/pdf/Defense_Program.pdf

———, *National Defense Program Guidelines for FY 2014 and Beyond*, Tokyo, December 17, 2013b. As of November 30, 2016:
http://www.mod.go.jp/j/approach/agenda/guideline/2014/pdf/20131217_e2.pdf

———, *Defense of Japan 2014*, Tokyo, August 6, 2014.

Ministry of Foreign Affairs of Japan, "Japanese Territory: Senkaku Islands—Situation of the Senkaku Islands," web page, April 4, 2014a. As of November 29, 2016:
http://www.mofa.go.jp/a_o/c_m1/senkaku/page1we_000010.html

———, "Japanese Territory: Senkaku Islands," web page, April 13, 2014b. As of November 29, 2016:
http://www.mofa.go.jp/region/asia-paci/senkaku/index.html

———, "Cabinet Decision on Development of Seamless Security Legislation to Ensure Japan's Survival and Protect Its People," web page, July 1, 2014c. As of December 1, 2016:
http://www.mofa.go.jp/fp/nsp/page23e_000273.html

———, *The Guidelines for Japan-U.S. Defense Cooperation*, Tokyo, April 27, 2015a. As of February 10, 2017:
http://www.mod.go.jp/e/d_act/anpo/shishin_20150427e.html

———, *Defense of Japan 2016*, Tokyo, December 2015b. As of February 24, 2017:
http://www.mod.go.jp/e/d_budget/pdf/280330.pdf

Ministry of National Defense of the People's Republic of China, *2013 Republic of China National Defense Report*, Beijing, October 8, 2013.

———, "Defense Ministry's Regular Press Conference on June 30," transcript, June 30, 2016. As of November 30, 2016:
http://eng.mod.gov.cn/Press/2016-06/30/content_4685792.htm

Ministry of National Defense of the Republic of China, *2013 Quadrennial Defense Review*, Taipei, March 2013a.

———, *National Defense Report: 2013*, Taipei, October 2013b.

Minnick, Wendell, "Taiwan Working on New 'Cloud Peak' Missile," *DefenseNews*, January 18, 2013.

Miranda, Mark, "Artillery, Air Crews Execute HI-RAIN Joint Exercise," U.S. Army, February 6, 2014. As of February 15, 2017:
https://www.army.mil/article/119731/
Artillery__air_crews_execute_HI_RAIN_joint_exercise

Mirasola, Chris, "Water Wars: The Philippines Pivots Towards Beijing," *Lawfare*, September 30, 2016. As of November 30, 2016:
https://www.lawfareblog.com/water-wars-philippines-pivots-towards-beijing

Moseley, T. Michael, *Operation IRAQI FREEDOM—By the Numbers: Assessment and Analysis Division*, Shaw Air Force Base, S.C.: U.S. Central Command Air Forces, April 30, 2003.

Naval Today, "Poland Orders NSM from Kongsberg," December 19, 2014. As of October 8, 2016:
http://navaltoday.com/2014/12/19/poland-orders-nsm-from-kongsberg/

North Atlantic Treaty Organization, "Joint Press Point—with NATO Deputy Secretary General Ambassador Alexander Vershbow and Italian Undersecretary of Defence, Gioacchino Alfano," transcript, last updated October 20, 2015. As of November 23, 2016:
http://www.nato.int/cps/en/natohq/opinions_124025.htm

———, "'NATO Post-Warsaw: Strengthening Security in a Tough Neighbourhood'—Speech by NATO Deputy Secretary General Ambassador Alexander Vershbow at the Annual Meeting of Romanian Ambassadors in Bucharest," transcript, August 29, 2016. As of November 23, 2016:
http://www.nato.int/cps/en/natohq/opinions_134512.htm?selectedLocale=en

Obama, Barack, *National Security Strategy*, Washington, D.C.: The White House, May 2010. As of February 5, 2017:
http://nssarchive.us/NSSR/2010.pdf

———, *National Security Strategy*, Washington, D.C.: The White House, February 2015. As of February 5, 2017:
http://nssarchive.us/wp-content/uploads/2015/02/2015.pdf

O'Connor, Sean, "PLA Ballistic Missiles," Air Power Australia, January 27, 2014. As of December 1, 2016:
http://www.ausairpower.net/APA-PLA-Ballistic-Missiles.html#mozTocId8319

Office of the Chief of Naval Operations, *Report to Congress on the Annual Long-Range Plan for Construction of Naval Vessels for FY2015*, Washington, D.C.: U.S. Department of Defense, June 2014. As of February 15, 2017:
http://navylive.dodlive.mil/files/2014/07/30-year-shipbuilding-plan1.pdf

Office of the Director of National Intelligence, "Global Trends 2030: Alternative Worlds," web page, 2012. As of November 30, 2016:
http://www.dni.gov/index.php/about/organization/
national-intelligence-council-global-trends

Office of the Secretary of Defense, *Annual Report to Congress: Military and Security Developments Involving the People's Republic of China 2014*, Washington, D.C.: U.S. Department of Defense, E-6A4286B, April 2014.

O'Rourke, Ronald, *China Naval Modernization: Implications for U.S. Navy Capabilities—Background and Issues for Congress*, Washington, D.C.: Congressional Research Service, September 8, 2014.

Permanent Court of Arbitration, "Summary of the Tribunal's Decisions on Its Jurisdiction and on the Merits of the Philippines's Claims," in *The South China Sea Arbitration (The Republic of the Philippines v. The People's Republic of China)*, The Hague, July 12, 2016. As of November 23, 2016:
https://pca-cpa.org/wp-content/uploads/sites/175/2016/07/
PH-CN-20160712-Press-Release-No-11-English.pdf

Pew Research Center, *Global Opposition to U.S. Surveillance and Drones, but Limited Harm to America's Image: Many in Asia Worry About Conflict with China*, Washington, D.C., July 14, 2014. As of November 30, 2016:
http://www.pewglobal.org/files/2014/07/2014-07-14-Balance-of-Power.pdf

Philips, Tom, "Taiwan's New President Tsai Ing-wen Vows to Reduce Dependence on Beijing," *Guardian*, May 20, 2016. As of November 23, 2016:
https://www.theguardian.com/world/2016/may/20/
taiwans-new-president-tsai-ing-wen-vows-to-reduce-dependence-on-beijing

Prins, Brandon, "Interstate Rivalry and the Recurrence of Crises: A Comparison of Rival and Nonrival Crisis Behavior, 1918–1994," *Armed Forces and Society*, April 1, 2005.

Public Law 96-8, Taiwan Relations Act, 96th Congress, Washington, D.C., January 1, 1979. As of November 23, 2016:
http://www.ait.org.tw/en/taiwan-relations-act.html

Quan Xianlian, "The Right to Reject Tribunal Ruling Is Real," *China Daily*, July 11, 2014.

Rapkin, David P., and William, R. Thompson, *Transition Scenarios: China and the United States in the Twenty-First Century*, Chicago: University of Chicago Press, 2013.

Rauhala, Emily, "Philippines Says China Has Stopped Chasing Fishermen from Contested Shoal," *Washington Post*, October 28, 2016. As of February 11, 2017:
https://www.washingtonpost.com/world/
report-filipino-fishermen-return-to-fish-shoal-contested-with-china/2016/10/28/
51d51eb4-9cb3-11e6-b4c9-391055ea9259_story.html?utm_term=.24f50c6937b5

Raytheon, "Naval Strike Missile: Adaptable, Long-Range Precision," 2016. As of December 1, 2016:
http://www.raytheon.com/capabilities/products/nsm/

Ryall, Julian, "Japan Agrees to Buy Senkaku Islands," *Telegraph*, September 5, 2012. As of November 29, 2016:
http://www.telegraph.co.uk/news/worldnews/asia/japan/9521793/
Japan-agrees-to-buy-disputed-Senkaku-islands.html

Santos, Matikas, "Chinese Ships Leave Paracel Islands After Landing Drills," *Inquirer*, January 23, 2014. As of November 30, 2016:
http://globalnation.inquirer.net/97521/
chinese-ships-leave-paracel-islands-after-landing-drills/

Schreer, Benjamin, "China's Growing Military Might Has Japan on Edge: Tokyo Responds," *The National Interest, The Buzz*, August 8, 2014. As of December 1, 2016:
http://nationalinterest.org/blog/the-buzz/
chinas-growing-military-might-has-japan-edge-tokyo-responds-11043

Schwartz, Stephen I., *Atomic Audit: The Costs and Consequences of U.S. Nuclear Weapons Since 1940*, Washington, D.C.: Brookings Institution Press, June 1, 1998.

Shan Jie, "China to Hold Drills Near Xisha Islands: Exercises in S. China Sea Not Linked to Arbitration: Expert," *Global Times*, July 4, 2016. As of November 29, 2016:
http://www.globaltimes.cn/content/992071.shtml

Shlapak, David A., and Michael Johnson, *Reinforcing Deterrence on NATO's Eastern Flank: Wargaming the Defense of the Baltics*, Santa Monica, Calif.: RAND Corporation, RR-1253-A, 2016. As of November 23, 2016:
http://www.rand.org/pubs/research_reports/RR1253.html

Shlapak, David A., David T. Orletsky, Toy I. Reid, Murray Scot Tanner, and Barry Wilson, *A Question of Balance Political Context and Military Aspects of the China-Taiwan Dispute*, Santa Monica, Calif.: RAND Corporation, MG-888-SRF, 2009. As of November 23, 2016:
http://www.rand.org/pubs/monographs/MG888.html

Sta. Ana, D. J., "China Reclaiming Land in Five Reefs?" *Philippine Star*, June 13, 2014. As of November 29, 2016:
http://www.abs-cbnnews.com/nation/06/13/14/China-reclaiming-land-5-reefs

State Council Information Office, *China's National Defense*, white paper, Beijing, 2010. As of November 30, 2016:
http://www.China.org.cn/english/features/book/194485.htm

———, *China's Peaceful Development*, white paper, Beijing, September 6, 2011. As of November 23, 2016:
http://english.gov.cn/archive/white_paper/2014/09/09/content_281474986284646.htm

———, *Diaoyu Dao, an Inherent Territory of China*, Beijing, September 24, 2012.

———, *The Diversified Employment of China's Armed Forces*, white paper, Beijing, April 2013. As of March 1, 2017:
http://aseanregionalforum.asean.org/files/library/ARF%20Defense%20White%20Papers/China-2013.pdf

Stillion, John, and David T. Orletsky, *Airbase Vulnerability to Conventional Cruise-Missile and Ballistic-Missile Attacks*, Santa Monica, Calif.: RAND Corporation, MR-1028-AF, 1999. As of November 23, 2016:
http://www.rand.org/pubs/monograph_reports/MR1028.html

Swaine, Michael, Mike M. Mochizuki, Michael L. Brown, Paul S. Giarra, Douglas H. Paal, Rachel Esplin Odell, Raymond Lu, Oliver Palmer, and Xu Ren, *China and the U.S.-Japanese Alliance in 2030: A Net Assessment*, Washington, D.C.: Carnegie Endowment for International Peace, May 3, 2013. As of November 30, 2016:
http://carnegieendowment.org/2013/05/03/
china-s-military-and-u.s.-japan-alliance-in-2030-strategic-net-assessment

Thayer, Carl, "Analyzing the US-Philippine Enhanced Defense Cooperation Agreement," *The Diplomat*, May 2, 2014.

Thomas, Brent, Mahyar Amouzegar, Rachel Costello, Robert A. Guffey, Andrew Karode, Christopher Lynch, Kristin Lynch, Ken Munson, Chad J. R. Ohlandt, Daniel M. Romano, Ricardo Sanchez, Robert S. Tripp, and Joseph Vesely, *Project AIR FORCE Modeling Capabilities for Support of Combat Operations in Denied Environments*, Santa Monica, Calif.: RAND Corporation, RR-427-AF, 2015. As of February 27, 2017:
http://www.rand.org/pubs/research_reports/RR427.html

U.S. Air Force, "MC-130E/H Combat Talon I/II," fact sheet, March 28, 2003. As of August 27, 2014:
http://www.af.mil/AboutUs/FactSheets/Display/tabid/224/Article/104534/mc-130eh-combat-talon-iii.aspx

U.S. Army, "Indirect Fire Protection Capability, Increment 2–Intercept (IFPC Inc 2-I)," web page, undated. As of January 2, 2015:
http://www.msl.army.mil/Pages/CMDS/ifpc2.html

———, *Procurement Programs: Committee Staff Procurement Backup Book; Fiscal Year (FY) 2010 Budget Estimates, Missile Procurement*, Washington D.C., May 2009.

U.S. Department of Defense, *National Defense Strategy*, Washington, D.C., June 2008. As of November 29, 2016:
http://www.defense.gov/Portals/1/Documents/pubs/2008NationalDefenseStrategy.pdf

———, *HIMARS*, Selected Acquisition Report, Washington, D.C., December 2011.

———, *Joint Operational Access Concept (JOAC)*, Version 1.0, Washington, D.C., January 2012a. As of November 23, 2016:
http://www.defense.gov/Portals/1/Documents/pubs/JOAC_Jan%202012_Signed.pdf

———, *Sustaining U.S. Global Leadership: Priorities for 21st Century Defense*, Washington, D.C., January 2012b. As of November 23, 2016:
http://archive.defense.gov/news/Defense_Strategic_Guidance.pdf

———, *GMLTS/GMLRS AW*, Selected Acquisition Report, Washington, D.C., December 2012c.

———, *Annual Aviation Inventory and Funding Plan: Fiscal Years (FY) 2014–2043*, Washington, D.C., May 2013. As of February 15, 2017:
http://breakingdefense.com/wp-content/uploads/sites/3/2013/06/DoD-Aircraft-Report-to-Congress-.pdf

———, *Annual Report to Congress: Military and Security Developments Involving the People's Republic of China*, Washington, D.C., 2014a. As of February 6, 2017:
https://www.defense.gov/Portals/1/Documents/pubs/2014_DoD_China_Report.pdf

———, *Quadrennial Defense Review 2014*, Washington, D.C., 2014b. As of November 23, 2016:
http://archive.defense.gov/pubs/2014_Quadrennial_Defense_Review.pdf

———, *Joint Air-to-Surface Standoff Missile (JASSM)*, Selected Acquisition Report, Washington, D.C., December 2015.

———, *Fiscal Year (FY) 2017 President's Budget Submission: Chemical and Biological Defense Program*, Vol. 4: *Research, Development, Test & Evaluation, Defense-Wide*, Washington, D.C., February 2016a.

———, *Fiscal Year (FY) 2017 President's Budget Submission: Army Justification Book of Research, Development, Test & Evaluation, Army RDT&E*, Vol. 2: *Budget Activity 5*, Washington, D.C., February 2016b.

U.S. Department of State, "U.S. Collective Security Agreements," web page, undated. As of November 30, 2016:
http://www.state.gov/s/l/treaty/collectivedefense/

———, Treaty Between the United States of America and the Union of Soviet Socialist Republics on the Elimination of their Intermediate-Range and Shorter-Range Missiles (INF Treaty), December 8, 1987. As of February 15, 2017:
https://www.state.gov/t/avc/trty/102360.htm

van Tol, Jan, Mark Gunzinger, Andrew F. Krepinevich, and Jim Thomas, *AirSea Battle: A Point-of-Departure Operational Concept*, Washington, D.C.: Center for Strategic and Budgetary Assessments, 2010.

Vasquez, John, *What Do We Know About War?* New York: Rowman and Littlefield, 2012.

Washington Post, "Q&A: Japan's Yomiuri Shimbun Interviews President Obama," April 23, 2014. As of February 6, 2017:
https://www.washingtonpost.com/world/qanda-japans-yomiuri-shimbun-interviews-president-obama/2014/04/23/d01bb5fc-cae3-11e3-95f7-7ecdde72d2ea_story.html?utm_term=.1765aaef072d

Weisgerber, Marcus, "Pentagon Debates Policy to Strengthen, Disperse Bases," *Defense News*, April 13, 2014.

White House, "Remarks by President Obama to the People of Estonia," Washington, D.C.: Office of the Press Secretary, September 3, 2014a. As of November 23, 2016:
https://www.whitehouse.gov/the-press-office/2014/09/03/remarks-president-obama-people-estonia

———, "Remarks by President Obama and President Xi Jinping in Joint Press Conference," Washington, D.C.: Office of the Press Secretary, November 12, 2014b. As of November 30, 2016:
http://www.whitehouse.gov/the-press-office/2014/11/12/remarks-president-obama-and-president-xi-jinping-joint-press-conference

———, "Remarks by President Obama and Prime Minister Abe of Japan in Joint Press Conference, Rose Garden," Washington, D.C.: Office of the Press Secretary, April 28, 2015.

Winnefeld, James A., Preston Niblack, and Dana J. Johnson, *A League of Airmen: U.S. Air Power in the Gulf War*, Santa Monica, Calif.: RAND Corporation, MR-343-AF, 1994. As of November 23, 2016:
http://www.rand.org/pubs/monograph_reports/MR343.html

Woolf, Amy F., *Conventional Prompt Global Strike and Long-Range Ballistic Missiles: Background and Issues*, Washington, D.C.: Congressional Research Service, R41464, February 24, 2016. As of December 15, 2016:
http://www.fas.org/sgp/crs/nuke/R41464.pdf

World Bank, *Global Economic Prospects: Divergences and Risks*, Washington, D.C., 2013. As of November 30, 2016:
http://www.worldbank.org/en/publication/global-economic-prospects

Xi Jinping, "New Asian Security Concept for New Progress in Security Cooperation," remarks at the Fourth Summit of the Conference on Interaction and Confidence Building Measures in Asia, Shanghai, May 21, 2014. As of February 24, 2017:
http://www.fmprc.gov.cn/mfa_eng/zxxx_662805/t1159951.shtml

Xinhua, "China Cracks Down on Erroneous Maps," January 9, 2013a.

———, "Xi Jinping Stresses at the Third Collective Study Session of the Political Bureau to Make Overall Planning for Domestic and International Situations," January 29, 2013b.

———, "The Diversified Employment of China's Armed Forces," April 19, 2013c.

———, "Xi Jinping's Remarks at the Fourth Conference on Interaction and Confidence Building Measures," May 21, 2014a.

———, "FM Spokesman Comments on Japan's Statement of Defense on Senkaku," November 22, 2014b.

———, "Full Text: CHINA Government Position Paper on Matter of Jurisdiction in South China Sea Arbitration Initiated by Philippines," December 7, 2014c.

———, "Commentary: Philippines Wise to Draw Experience from U.S. Bloody Record of Intervention," July 10, 2016.

Yao Jianing, "Chinese Navy to Conduct Drills Around Xisha Islands," *China Military Online*, July 6, 2016. As of February 24, 2017:
http://english.chinamil.com.cn/news-channels/china-military-news/2016-07/06/content_7138788.htm

Yu Jixun, ed., *The Science of Second Artillery Campaigns*, Beijing: PLA Press, 2004.

Zambo Times, "PH to Acquire Shore-Based Missile System," December 4, 2013.

Zhang Yuliang, ed., *The Science of Campaigns*, Beijing: National Defense University Press, 2006.

Zhang Yunbi, "China's Air Force Flags Regular Patrols in South China Sea," *China Daily*, July 19, 2016. As of February 3, 2017:
http://usa.chinadaily.com.cn/china/2016-07/18/content_26133352.htm

Lightning Source UK Ltd.
Milton Keynes UK
UKHW021300220721
387597UK00012B/108